MOUNTAIN PLANTS
OF THE
PACIFIC NORTHWEST

A Field Guide to Washington,
Western British Columbia, and Southeastern Alaska

RONALD J. TAYLOR

AND

GEORGE W. DOUGLAS

Illustrations by Gail F. Harc
Linda Vorobik, and Alice An

Mountain Press Publishing Company
Missoula, Montana
1995

Illustrations by Gail F. Harcombe,
Linda Vorobik, and Alice Anderson

Map by Ed Jenne

All photographs by Ronald J. Taylor unless otherwise noted.

Thanks to Kevin Short for his finesse with computers.

Library of Congress Cataloging-in-Publication Data

Taylor, Ronald J., 1932-
Mountain plants of the Pacific Northwest : a field guide to Washington, western British Columbia, and southeastern Alaska / Ronald J. Taylor and George W. Douglas ; illustrations by Gail F. Harcombe, with Linda Vorobik and Alice Anderson.
p. cm.
Includes bibliographical references (p.) and index.
ISBN 0-87842-314-1 (pbk.)
1. Mountain plants–Northwest, Pacific–Identification. 2. Mountain plants–Alaska–Identification. I. Douglas, George W. (George Wayne), 1938- . II. Title.
QK144.T35 1995 95-14240
581.9795–dc20 CIP

Printed in Hong Kong by Mantec Production Company

Mountain Press Publishing Company
P.O. Box 2399 • Missoula, MT 59806
406-728-1900 • 800-234-5308

To the many students with whom I have explored the floristically diverse and beautiful mountains of the Pacific Northwest.
–R. J. T.

To Dr. Arthur R. Kruckeberg,
whose patience and inspiriation during my first serious plant identification efforts started me on the road to a lifelong career in botoany.
–G. W. D.

Contents

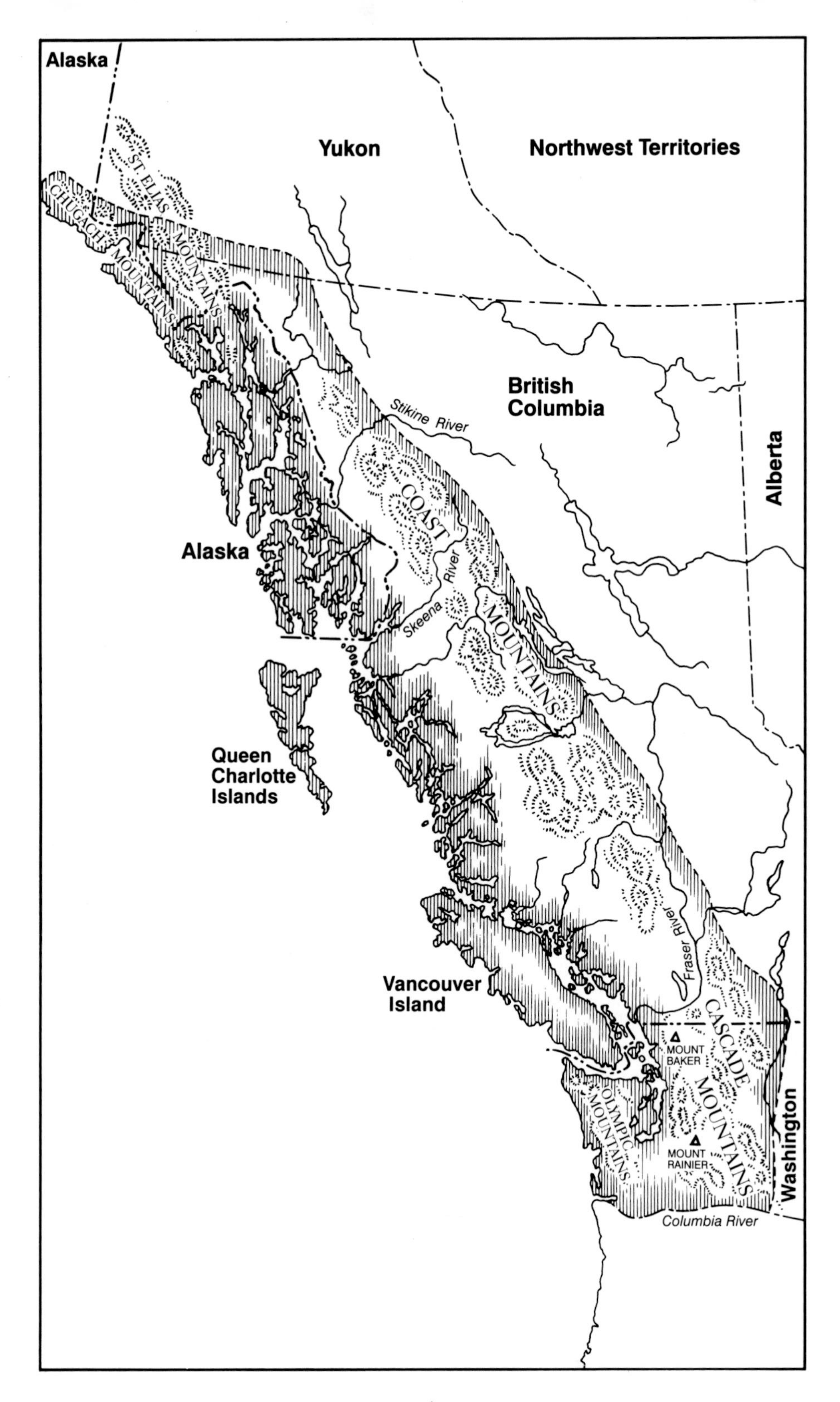
Alaska
Yukon
Northwest Territories
ST. ELIAS MOUNTAINS
CHUGACH MOUNTAINS
British Columbia
Stikine River
Alberta
COAST MOUNTAINS
Skeena River
Alaska
Queen Charlotte Islands
Fraser River
Vancouver Island
MOUNT BAKER
CASCADE MOUNTAINS
OLYMPIC MOUNTAINS
MOUNT RAINIER
Washington
Columbia River

Introduction

The mountains of Washington, western British Columbia, and southeastern Alaska form a natural system integrated by climate, vegetation, and geological history. The area is also one of nature's masterpieces. Beauty may be subjective, but anyone would be impressed by the grandeur of these lofty mountains–the striking color contrasts between the green forests, the multicolored wildflowers on the alpine and subalpine slopes, the brilliant white snowfields, and the pale blue ice fields. Add to this setting the many lakes and the sparkling streams cascading to the rivers below, and even the most casual observer would agree the setting is beautiful.

WESTERN SLOPES

Continuous Forest. Forests on the moist, western slopes of Pacific Northwest mountain ranges are famous for their giant trees, but visitors should be careful not to miss the forest for the trees. An abundance of epiphytic mosses and lichens cover tree trunks and hang from branches throughout the forest ecosystem. Varied understory mosses, ferns, herbs, and shrubs blend together in multiple shades of green in the soft light of the forest floor.

The total number of plant species in these western-slope forests is not particularly large, however, and vegetative continuity characterizes each distinct habitat. Ferns typically proliferate along streams, together with water-loving forbs such as members of the Saxifrage family. Many shrubs also thrive in stream habitats, often forming dense thickets. Among these are devil's club (a satanic shrub with vicious thorns), prickly currant, salmonberry, thimbleberry, and various willows. Where the forest is open and free of oppressive shade, vegetation is more diverse, and shrubs tend to dominate, typically vine maple, red elderberry, Indian plum, salmonberry, thimbleberry, and goat's beard. Members of the heath family (Ericaceae) grow well in both open and closed forests, especially huckleberries and blueberries (species of *Vaccinium*) and wintergreens *(Pyrola).*

Prominent among the other regular and conspicuous inhabitants of western continuous forests are the ferns. Ferns thrive everywhere and commonly form dense communities, especially sword fern, deer fern, and lady-fern. Among the most widely distributed and ecologically important forbs and low shrubs are foamflower, strawberry bramble, twinflower, bunchberry, queen's cup lily, and rosy twisted-stalk. Salal and Oregon grape are common understory dominants in the foothills, especially in relatively dry sites. Beautiful trillium and unusual wild bleedingheart are in the forefront of the advance from stark winter into vibrant summer. Yellow violets grow in colorful populations where the litter layer is poorly developed. Equally dense and attractive populations of wild lily-of-the-valley proliferate where the soil is moist and humus rich. Many other plants in the region, although less abundant, are conspicuous because of their unusual appearance. These include the non-green ericads (members of the heath family) and coral root orchids. These plants can grow in deep shade, since they derive their energy from associated fungi rather than the sun.

Contrary to what many people imagine, the continuous forest is not by itself a suitable habitat for most well-known wildlife of the Northwest. Exceptions are the ruffed grouse, spotted owl, tree squirrels, chipmunks, and the elusive marten and weasel. Birds and large mammals use the forest for shelter or food but spend much of their life elsewhere, such as the scattered tree clumps and meadows of the subalpine region. Along waterways and lakes, the fauna is more diverse and includes the majestic bald eagle.

To simplify study and discussion, different forest zones are given names that relate to elevation and climax dominant species. The immediate coastal area is called the **Sitka spruce zone.** Although Sitka spruce extends inward and upward, it does so as a seral species, ultimately replaced by more shade-

Sitka spruce community with devil's club

tolerant climax species. Inland from the coast, the **western hemlock zone** extends upward about 3,000 feet (914 m), less so to the north. Western red cedar is often a codominant with western hemlock at lower elevations and along drainages, and Pacific silver fir is a codominant at upper elevations in Washington and southern British Columbia. The **mountain hemlock zone** extends upward from the western hemlock zone to the treeline. In northern

Western hemlock and silver fir crowns

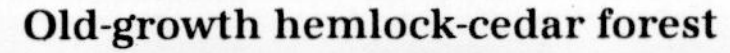

Old-growth hemlock-cedar forest

British Columbia and southeastern Alaska, white spruce is a codominant with (or often replaces) mountain hemlock, particularly at upper elevations. On western slopes, western hemlock and red cedar forests dominate the **lower montane zone,** and mountain hemlock and silver fir forests dominate the **upper montane zone.**

Subalpine Zone. The subalpine zone is easier to describe than define. It is characterized by scattered tree clumps in a meadow mosaic. Although the prevailing climate is sufficiently mild to allow upright tree growth, other factors prevent the development of a continuous, or closed, forest. On some slopes, the discontinuity of the forest is the result of periodic avalanches. The vegetation is diverse along avalanche tracks and includes shrubs normally associated with open forests–salmonberry, thimbleberry, elderberry, mountain ash, Sitka alder, willows, and huckleberries–and tall herbs typical of mountain meadows–cow parsnip, edible thistle, tiger lily, northern riceroot, false hellebore, Sitka valerian, broadleaf lupine, Fendler's waterleaf, western meadowrue, and columbines. Other factors, singly or in combination, promote the discontinuity of the forest and the formation of meadows in the subalpine zone: high winds; late snowmelt and a resulting short season; summer drought, especially on south-facing slopes; competition between tree seedlings and meadow species; and inadequate soils, especially on ridgetops and rock outcrops.

In some areas, trees are invading the meadows. If one tree, such as a mountain hemlock, subalpine fir, silver fir, or white spruce, becomes established in

Autumn subalpine scene in the North Cascades

Subalpine meadow in northwestern British Columbia —George W. Douglas photo

a meadow, it alters the immediate environment and enables other tree seedlings to grow. In this way, tree clumps develop and have the potential ultimately to expand and form a closed forest. These subalpine trees are normally short, with spreading branches, but retain definite crowns and do not develop the dense, low, thicketlike growth form known as krummholz. Shrubs prevail in the immediate vicinity of subalpine tree clumps, and commonly include white rhododendron, maple-leaf currant, false azalea, and huckleberries.

Vegetation in the open meadows is lush and diverse, including several sedges and grasses and a wide variety of colorful herbaceous flowering plants: mountain bistort, asters, lupines, louseworts, fanleaf cinquefoil, Sitka valerian, paintbrushes, partridge foot, broad-leaf arnica, mountain daisy, and the well-known early flowering glacier lily, spring beauty, and western anemone. In moist but well-drained areas, various heathers and huckleberries are especially common. Colorful wildflowers such as spreading phlox, Davidson's penstemon, stonecrops, and kinnikinnick prevail on rock outcrops. Attractive riparian communities thrive along waterways and in seepage areas, dominated by species such as showy sedge, red monkeyflower, bog orchids, saxifrages, showy gentian, marsh marigold, and willows.

The popularity of the subalpine zone easily exceeds that of other zones. The meadows are lush and strikingly attractive, with an assortment of colorful wildflowers. Wild animals regularly move between forest and meadow or feed on lush meadow vegetation. Blue grouse are often heard (but seldom seen) as they emit low basal calls from the shelter of tree clumps. The shrill whistle of marmots pierces the thin mountain air as sentinels warn of approaching danger. Marmot or Arctic ground squirrel burrows characteristically scar the meadows, especially in dense vegetation below snowfields. Pikas are common inhabitants of coarse, talus slopes and rock outcrops. Deer frequent the region, grazing on meadow vegetation or browsing on shrubs in and around tree clumps, but never wandering far from the shelter of the continuous forest. Mountain goats and Dall sheep (in northern British Columbia and southeastern Alaska) range from the upper limits of the forest zone to alpine slopes above. Bears forage throughout the area, especially in late summer and autumn when huckleberries are ripe. Several carnivores, including the majestic cougar, frequent the region but are rarely seen. Finally, evidence of small rodents is everywhere, the most obvious being the latrine piles, long dirt mounds, and nests of dried plants left behind by the heather vole after a winter of burrowing beneath the snow.

Alpine Zone. At higher elevations, trees become further dwarfed, forming the characteristic krummholz stands of mountain hemlock and subalpine fir, or white spruce in the north. Only those branches low enough to the ground to be sheltered by snow escape the scarring effects of freezing temperatures, high winds, and winter drought. Krummholz formations mark the beginning of the alpine zone, which is a land of contrasts. Temperatures range from well below freezing in the winter to as much as 90 degrees Fahrenheit (32° C)

Subalpine fir–krummholz

in August. Even during summer months, freezing temperatures are routine, with extreme temperature variations between day and night or from one day to the next. Winds, which are typically strong, may have an extreme chilling effect or may provide relief from the burning sun. Even on the western side of the mountain crests, summers are dry, and drought conditions normally develop on well-drained, south-facing slopes. In contrast, on northern slopes snow accumulation is much greater, and the snow melts later. Here, the limiting factor for plant growth is cool temperatures and a short season, not lack of sufficient moisture.

The soil of alpine slopes is unstable, poorly developed, and easily eroded. Frost action uproots plants and loosens the soil, increasing the rate of erosion. Gravel stripes mark the pattern of water percolation down the slopes and result in vegetation-stripe communities. Turf mats stand above scarred terrain, providing stark evidence of the rate and extent of erosion. On some moist slopes the soil itself "flows" downward, imperceptibly slowly, looking like waves of water from a distance.

Within this hostile environment, plants must have special adaptations to survive. They must be generalists, tolerating extreme and variable climates and adverse soil conditions. Many plants have a low, cushion form, which provides protection from wind and desiccation and concentrates solar energy at ground level, warming the roots. Most alpine plants are covered with silky hair, which insulates them from rapid temperature change and reduces water loss on hot, windy summer days. Animals that live in or frequent the alpine zone must be hardy, but unlike plants, they can seek shelter elsewhere during adverse conditions.

Plant communities of the alpine zone vary according to soil conditions, water availability, and geographic location. In depressed areas, poor drainage results in boglike sedge meadows. In well-developed soil, vegetation cover is more or less continuous and meadowlike, with grasses and sedges sharing dominance with an assortment of forbs and low shrubs such as dwarfed heathers, willows, and huckleberries. But most of the alpine region consists of talus (rocky) slopes, fellfields (gravelly ridgetops), rock formations, or glaciers and snowfields. In fine talus, fellfields, and on rock outcrops and ledges, the vegetation is discontinuous, consisting of low, typically matted plants, many with beautiful flowers. Some common species of these alpine habitats are alpine pussytoes, moss campion, silky phacelia, daisies, shrubby cinquefoil, spreading phlox, stonecrops, penstemons, wild buckwheats, saxifrages, cliff paintbrush, and several species of the pink and mustard families. Attractive and often bright-colored lichens comprise an important part of alpine ecosystems, some appearing to be painted on the rocks.

Although many animals spend part of their life in the alpine zone, few live there year-round. The animals most often associated with the high, treeless slopes are mountain goat, Dall sheep, and ptarmigan, the last as fearless as the first two are wary. The ptarmigan is a master of camouflage, its plumage white in winter and gray-brown in summer. Even during the molting season, its white and brownish coloration matches the surrounding boulder world.

EASTERN SLOPES

On the eastern mountain slopes, the environment is strongly influenced by a drier climate, the result of the rain-shadow effect. Plant communities here more closely resemble those of the Rocky Mountains. Two particularly common and conspicuous mat-forming shrubs rarely found on the western side of the crests are mountain avens and crowberry. Krummholz species are Engelmann spruce, subalpine fir, mountain hemlock (in particularly moist areas), and white spruce (in the north). Two additional resident trees low in stature but not truly krummholz are whitebark pine and subalpine larch. This same combination of trees makes up the tree clumps in the subalpine zone. Subalpine meadows are generally not as lush as on the moister, western side but are more diverse because of the influx of Rocky Mountain species and the infusion of grassland–sagebrush steppe species such as asters, buckwheats, paintbrushes, desert parsleys, and various grasses.

In Washington and southern British Columbia, Ponderosa pine and Douglas fir forests occupy the **lower montane zone,** and Engelmann spruce–subalpine fir forests occupy the **upper montane zone.** These two species also extend upward into the **krummholz zone** and downward, as seral species, into the Douglas fir zone. The **Douglas fir zone** is broad, with a heteroge-

Engelmann spruce–subalpine fir zone, subalpine larch in autumn foliage in background

Mixed conifer forest—larch in autumn foliage

neous mix of associated trees—lodgepole pine, western larch, western white pine, grand fir, and quaking aspen. Even though it is better adapted and grows larger on the western slopes, Douglas fir is primarily a seral species on the western slopes and a climax species on the eastern slopes. This is because hemlocks and firs, which are more shade tolerant and replace Douglas fir on the western slopes, are not sufficiently drought tolerant to grow on the eastern slopes.

The **ponderosa pine zone** is between the Douglas fir zone and the grassland–sagebrush steppe in Washington and southern British Columbia. Ponderosa pine depends on a combination of dry conditions and fire for its dominance over Douglas fir.

The Engelmann spruce–subalpine fir and Douglas fir zones are replaced, respectively, by an **Engelmann spruce–white spruce hybrid zone** at about 51 and 54 degrees north latitude. In northern British Columbia and southeastern Alaska, the Engelmann spruce influence fades, and white spruce grows in pure stands, constituting the **white spruce zone.** Paper birch or balsam poplar, sometimes both, form mixed stands with spruce.

Variations in moisture and temperature, largely associated with slope effect, will disrupt and shift zonation patterns. For example, western slope climax species spill over onto eastern slopes in areas where the climate is sufficiently moist.

Common shrubs in the drier, eastern forests include serviceberry, thimbleberry, birch-leaf spiraea, Douglas maple, mountain box, grouseberry, huckle-

Subalpine meadow in the North Cascades

berries, and wild roses. Among the most colorful wildflowers are heart-leaf arnica, Oregon anemone, star-flowered Solomon's seal, asters, lupines, wild geraniums, and ballhead waterleaf. In the open, low-elevation ponderosa pine forests and along dry, south-facing slopes, the equally attractive wildflowers of the sagebrush steppe replace most of the mountain plants. The colorful steppe species, particularly arrowhead balsamroot, put on a spectacular display. But the life and times of sagebrush country is another story.

HOW TO USE THIS GUIDE

This book is designed and written for anyone interested in the mountain plants of Washington, western British Columbia, and southeastern Alaska. We have attempted to restrict technical jargon, taking care not to sacrifice accuracy for simplicity. Also, to facilitate use of the book, descriptions of representative plants are systematically organized, accompanied by color photographs. In many cases, line drawings have been provided to illustrate technical terms and plant characteristics. Descriptions include information relating to ecology and distribution, along with notes of general interest.

The plants described in this book are in four categories: ferns and fern allies, trees, forbs and shrubs, and graminoids. Each group has major distinguishing characteristics.

Ferns and Fern Allies. Plants in this group are herbaceous and do not produce flowers or seeds but reproduce by spores. Fern leaves are all divided, while those of fern allies are very small and often scalelike.

Trees. Trees are woody plants that have a single stem (trunk) and grow to be several feet tall, sometimes hundreds of feet tall. Trees in the Pacific Northwest are either **conifers,** which are mainly evergreen and have needlelike or scalelike leaves, and **hardwoods,** the broad-leaf angiosperms. Conifers lack flowers and produce seeds mainly in woody cones. Angiosperms have flowers (although these are often highly reduced and inconspicuous) and produce seeds in some kind of fruit; for example, a nut, capsule, or berry.

Forbs and Shrubs. Forbs are herbaceous plants, other than the ferns and graminoids, and shrubs are woody plants with multiple stems. This large group of angiosperms is characterized, in most cases, by conspicuous flowers. The leaves of the group are extremely variable both in size and form but are seldom grasslike.

Graminoids. Graminoids are, by definition, grasslike plants. The leaves are long and narrow, with parallel veins, and the flowers are highly reduced and inconspicuous, adapted to wind pollination.

Ponderosa pine forest with balsamroot

Chowder Ridge and Mt. Shuksan, North Cascades –Anne Martin photo

A key has been provided for each group, located at the back of the book, to aid in identification of representative plants. Within each group, plants have been arranged alphabetically by family based on common names. Within the families, arrangement is alphabetical by scientific name (with some exceptions of convenience), since the common names of related species are typically unrelated. Common names of species are highlighted. The floristic treatment in some groups, trees for example, is much more complete than that of other groups. However, we have tried to include the most prevalent plants. A discussion of every plant that grows within the broad geographical range covered by this book would be impractical, if not impossible.

Plant nomenclature used in this book follows that of the *Flora of the Pacific Northwest,* by Hitchcock and Cronquist (1973), and/or the *Vascular Plants of British Columbia,* by Douglas et al. (1989–1994), except in cases where more recent name changes have been proposed. The common names have been chosen on the basis of general usage. Some terms used to describe specific species or plant parts are explained in the opening section for that family or genus, or in the Glossary. Labeled illustrations of important plant parts have been provided, along with appropriate indexes.

Measurements are given in roughly equivalent English and metric units. Those in English units are approximate; those in metric units are precise.

FERNS AND FERN ALLIES

The ferns and their allies are vascular plants; that is, they have conducting (vascular) tissues to transport water, nutrients, and various organic materials, especially sugars, throughout the plant. In this respect, they are related to conifers and flowering plants. They differ from those more advanced plants in reproducing by spores rather than seeds. Also, they have a more involved—though more primitive—life cycle, which includes independent gametophyte and sporophyte generations. The gametophyte is a small, inconspicuous plant, not much larger than a match head, that produces gametes, sperm, or eggs, or sometimes all. To effect fertilization, the sperm must swim through moisture in the soil to the egg on the same or different gametophyte. When the egg has been fertilized, it grows into an embryo and then into a mature spore-producing sporophyte, the dominant generation that we see and recognize. The spores are released into the wind, and when one falls to the ground, under suitable conditions it will develop into a gametophyte, completing the life cycle.

The ferns and fern allies comprise three major plant groups—ferns, clubmosses, and horsetails, which are easily differentiated. A few scientific terms relating to these plants are used in the plant descriptions. For people unfamiliar with the terms, a few moments digesting the definitions will enhance understanding of the text.

Clubmosses have leaves that are small and scalelike or lance shaped, never more than a few centimeters long, and are alternate or opposite around erect or spreading stems. Spores are produced in spore sacs on the upper (inner) surface of the leaves. In most species, the spore leaves are smaller than the other leaves and are usually densely clustered at the tip of erect branches, forming a strobilus, or simply a cone.

Horsetails are easily recognized by their hollow, jointed stems. Branches grow in whorls at the joints, and small leaves sheath the stem above the branches. Some horsetails are dimorphic; that is, they have two forms, one nonbranched, nongreen, and spore-producing and the other green and branched.

Ferns have large, divided leaves (fronds) that grow from poorly developed underground stems. Spores are produced in spore sacs (sporangia) clustered on the undersurface of leaf segments or along the leaf margins in an area called a sorus.

Clubmosses

Lycopodiaceae and Selaginellaceae

Clubmosses, or lycopods, are the most primitive of living vascular plants, derived from ancestors dating back to the Lower Devonian Period, some 400 million years ago. Tree forms existed during the Carboniferous Period, 300 million years ago, probably living in swamps as dominant members of the coal swamp flora. The poorly developed roots were probably bathed in water, and the crowns flourished in the high-humidity atmosphere. Perhaps a drying trend toward the end of the Carboniferous Period led to the extinction of the tree forms. In any case, today's clubmosses are all small, low-growing, mosslike plants, never more than 1 foot (30 cm) high.

Clubmosses are classified into three genera: ***Huperzia, Lycopodium*** (Lycopodiaceae), and ***Selaginella*** (Selaginellaceae), the **little-clubmosses** or **spikemosses.** The little-clubmosses are mat-forming plants with inconspicuous, quadrangular, nonstalked cones and small, scalelike leaves. *Huperzia* and *Lycopodium* species usually have elongate, branched stems that spread over the ground, with roots growing out of the lower side. Spore-bearing branches stand erect, allowing easy wind dispersal of the spores. The leaves are lance shaped or linear, and the cones are cylindrical in *Lycopodium* and not differentiated in *Huperzia.* The most common clubmoss, *Lycopodium clavatum,* typifies this growth form. The leaves of all three genera are evergreen.

Staghorn Clubmoss ***Lycopodium clavatum***

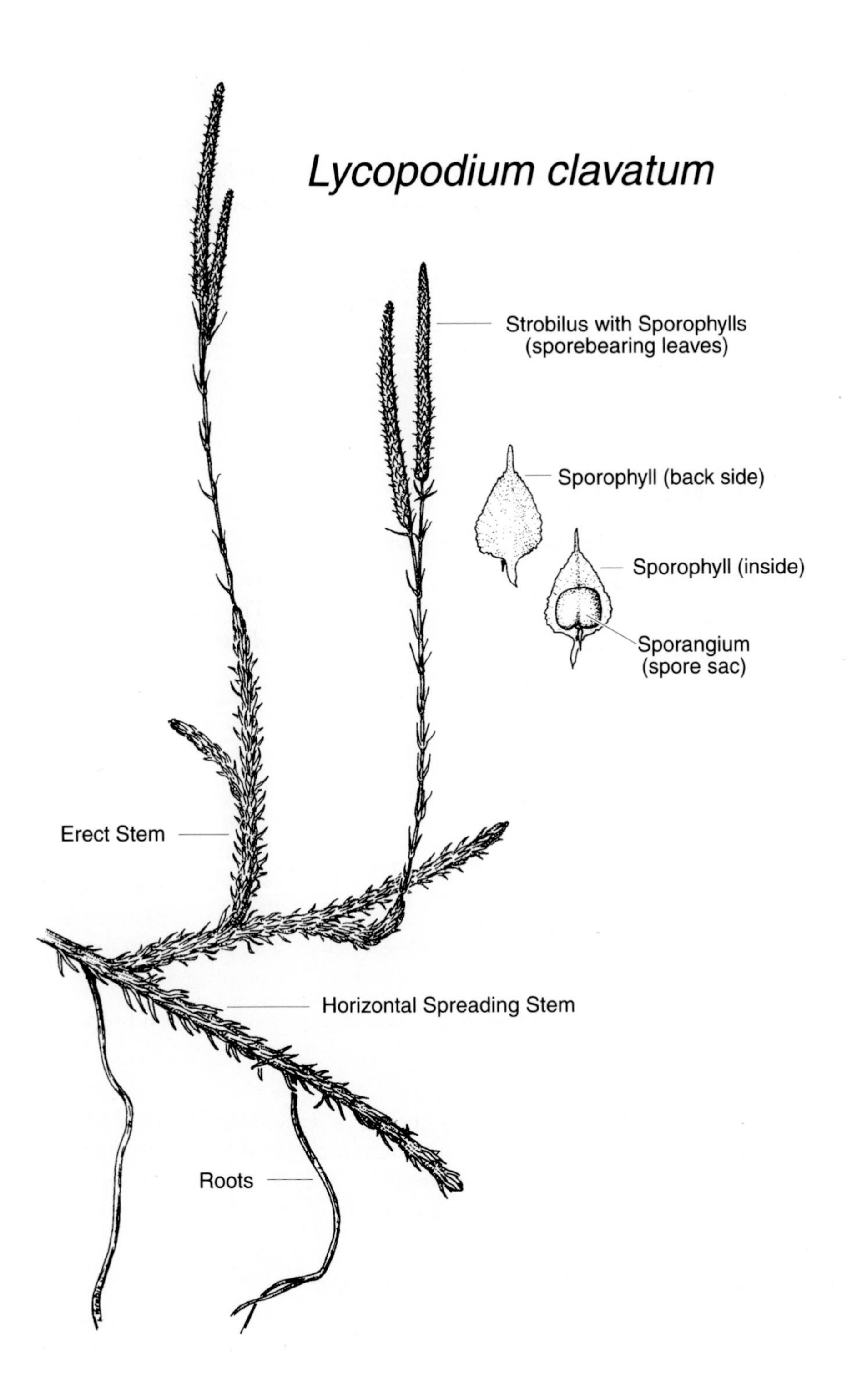
Lycopodium clavatum
Strobilus with Sporophylls
(sporebearing leaves)
Sporophyll (back side)
Sporophyll (inside)
Sporangium
(spore sac)
Erect Stem
Horizontal Spreading Stem
Roots

Fir Clubmoss

Huperzia (Lycopodium) selago

Fir clubmoss is unusual in that it does not produce a differentiated cone, and unlike most clubmosses, the horizontal stems of this species are short or absent. Erect stems may be branched or unbranched, often forming flat-topped tufts, and are usually less than 8 inches (20 cm) high. Roots develop at the base of the erect stems.

Leaves: Lance shaped to linear, spreading or reflexed backward (downward), tending to be lustrous.

Cones: Spore sacs are borne at the base of undifferentiated leaves in the upper part of the erect branches; that is, spore-bearing leaves resemble lower leaves rather than being modified into a cone.

Ecology: Common in boggy soils of the Arctic tundra and in wet, acidic soils of open forests at various elevations, more or less throughout the Pacific Northwest.

Other clubmoss species may easily be mistaken for fir clubmoss when they are in a nonfertile stage.

Stiff Clubmoss

Lycopodium annotinum

The horizontal, trailing stems of this species are usually less than 3 feet (1 m) long, are mainly unbranched, and root freely. Erect stems vary from 4 to 12 inches (10 to 30 cm) high and are usually branched. The plant often forms very dense populations.

Leaves: Lance shaped to linear, stiffly spreading, sharp pointed but not spiny.

Cones: Nonstalked, at the end of the leafy stem.

Ecology: Widely distributed but infrequently observed; common in Alaska and northern British Columbia, extending southward through the mountains of Washington into Oregon; prefers the wet, acidic soils of open montane forests, extending up to the subalpine zone.

Stiff clubmoss can be easily distinguished from other clubmosses by its erect, usually unbranched stems, with stiffly spreading leaves.

Fir Clubmoss *Huperzia selago*

Stiff Clubmoss *Lycopodium annotinum*

Stiff Clubmoss *Lycopodium annotinum*

Staghorn Clubmoss *Lycopodium clavatum*

Also called running clubmoss, the horizontal, trailing stems of this species are long and usually branched, forming a network over the forest floor and across decaying logs, rooting intermittently. The erect stems are about 1 foot (30 cm) high and often branched.

Leaves: Lance shaped with a threadlike tip, projected forward on the stem and often curved inward.

Cones: Borne on long, sparsely leafy stalks that are usually branched toward the tip, thus bearing two or more cones per stalk; the stalks with cones fancifully resemble stags' horns.

Ecology: Distributed throughout the Pacific Northwest in a variety of habitats, from bogs to dry rocky areas and from shady forest floors to open places, such as logged areas and road cuts, and from lowlands to the alpine zone.

Ground Cedar *Lycopodium complanatum*

This clubmoss is differentiated from all others by its flattened, extensively branched, cedarlike stems. The trailing stems form a network over the ground, intermittently rooting.

Leaves: Scalelike with a bristle tip, opposite in four rows, similar to the leaves of Pacific red cedar, thus the common name.

Cones: Usually borne in pairs on long, scaly stalks.

Ecology: Distributed more or less throughout the region but usually goes unobserved; grows in forests or forest openings, from lowlands to mid-elevations in the mountains.

Ground Pine *Lycopodium dendroideum (obscurum)*

This unusual clubmoss resembles a 1-foot (30 cm) pine tree, with numerous symmetrically spreading branches. The horizontal stems are subterranean rather than trailing over the ground.

Leaves: Lance shaped to linear with a sharp but not spiny tip, projected outward and forward at about 45 degrees.

Cones: Nonstalked, occurring at the end of erect, leafy branches, one cone per fertile branch.

Ecology: Common in western British Columbia but rare in Washington and Alaska; prefers moist, acidic soils in open forests at mid-elevations.

Staghorn Clubmoss *Lycopodium clavatum*

Ground Pine *Lycopodium dendroideum*

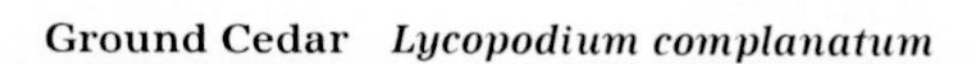

Ground Cedar *Lycopodium complanatum*

Sitka Clubmoss *Lycopodium sitchense*

The horizontal stems on Sitka clubmoss are often hidden by associated vegetation; rooting occurs intermittently. The many erect stems are extensively branched and usually less than 6 inches (15 cm) high, giving the plants a compact, tufted appearance.

Leaves: Small and lance shaped to scalelike, usually flattened against the stem.

Cones: Borne at the tips of leafy stalks that are not well differentiated from sterile branches, except the leaves lie flatter against the stem.

Ecology: A high-elevation plant, usually growing in alpine and subalpine meadows; common in Washington and British Columbia, less so in Alaska.

High-elevation forms of Sitka clubmoss are easily confused with the little-clubmosses, which are also low, dense, and cushionlike. However, Sitka clubmoss has thicker stems and much more distinctive cones.

Alpine Clubmoss *Lycopodium alpinum*

Horizontal stems of alpine clubmoss creep along the ground surface, rooting intermittently. Erect stems are 3 to 4 inches (7 to 10 cm) high and extensively branched, so the plants are densely tufted.

Leaves: Borne in four rows; the two opposite rows are distinct, the leaves of one row being lance shaped, those of the other row more scalelike, flattened against and partially fused to the stems.

Cones: Stalkless on short, leafy stems.

Ecology: Primarily an alpine species, as the common name suggests, but may extend downward along rocky ridges into open forests; distributed from Alaska southward to the North Cascades of Washington.

Alpine clubmoss may be confused with Sitka clubmoss, which grows in similar habitats. However, alpine clubmoss has scalelike opposite leaves alternating with lance-shaped opposite leaves, and the plants are blue-green, compared to the green or yellow-green Sitka clubmoss. The branches are flattened somewhat, in this respect bearing a resemblance to ground cedar *(L. complanatum).* Alpine clubmoss is rare south of British Columbia.

Sitka Clubmoss ***Lycopodium sitchense***

Sitka Clubmoss ***Lycopodium sitchense***

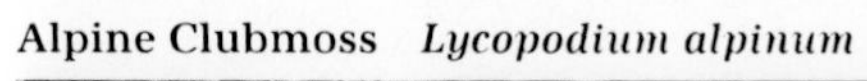

Alpine Clubmoss ***Lycopodium alpinum***

Wallace's Little-Clubmoss *Selaginella wallacei*

The plants are low, only about an inch high, and densely tufted. Horizontal stems are poorly developed, with threadlike roots arising from slender, lateral branches.

Leaves: Small, no more than ¼ inch (5 mm) long, scalelike and numerous, usually tipped by a tiny bristle.

Cones: Nonstalked and rather indistinct, but recognizable by their tightly compressed leaves and rectangular shape.

Ecology: This species of *Selaginella* ranges from southern British Columbia through Washington, growing mainly in exposed, rocky sites from lowlands to mid-elevations in the mountains.

Several other species of *Selaginella* grow in the Pacific Northwest. The most common is *S. densa,* a compact, mainly alpine plant with somewhat flattened branches. All species grow in rocky sites where soil development is poor, thereby largely avoiding competition with other plants. Such sites, however, become dry early in the season, subjecting the plants to extreme drought. The little-clubmosses have the unusual ability to survive desiccation, entering a period of dormancy until they become rehydrated and rejuvenated.

Wallace's Little-Clubmoss *Selaginella wallacei*

Horsetails (Horsetail Family) Equisetaceae

Like the clubmosses, the horsetails are primitive plants with treelike ancestors from the coal swamp flora of the Carboniferous Period. Existing horsetails are herbaceous (nonwoody) plants with underground rhizomes and erect stems. In most species, the stems are all alike, evergreen and spore bearing. In a few species, the stems are dimorphic; that is, they have two forms: green, branched, and sterile or nongreen (brownish), nonbranched, and spore bearing. The evergreen species usually grow along streams or around lakes. The leaves of horsetails are small, scalelike, and fused at the base to form a sheath around the jointed stem. The branches spread out in whorls. Cones are borne at the tip of the fertile stems.

Horsetails are also called **scouring rushes,** because of the rushlike appearance of the evergreen species and the silica embedded in the epidermis, which gives the plants an abrasive quality.

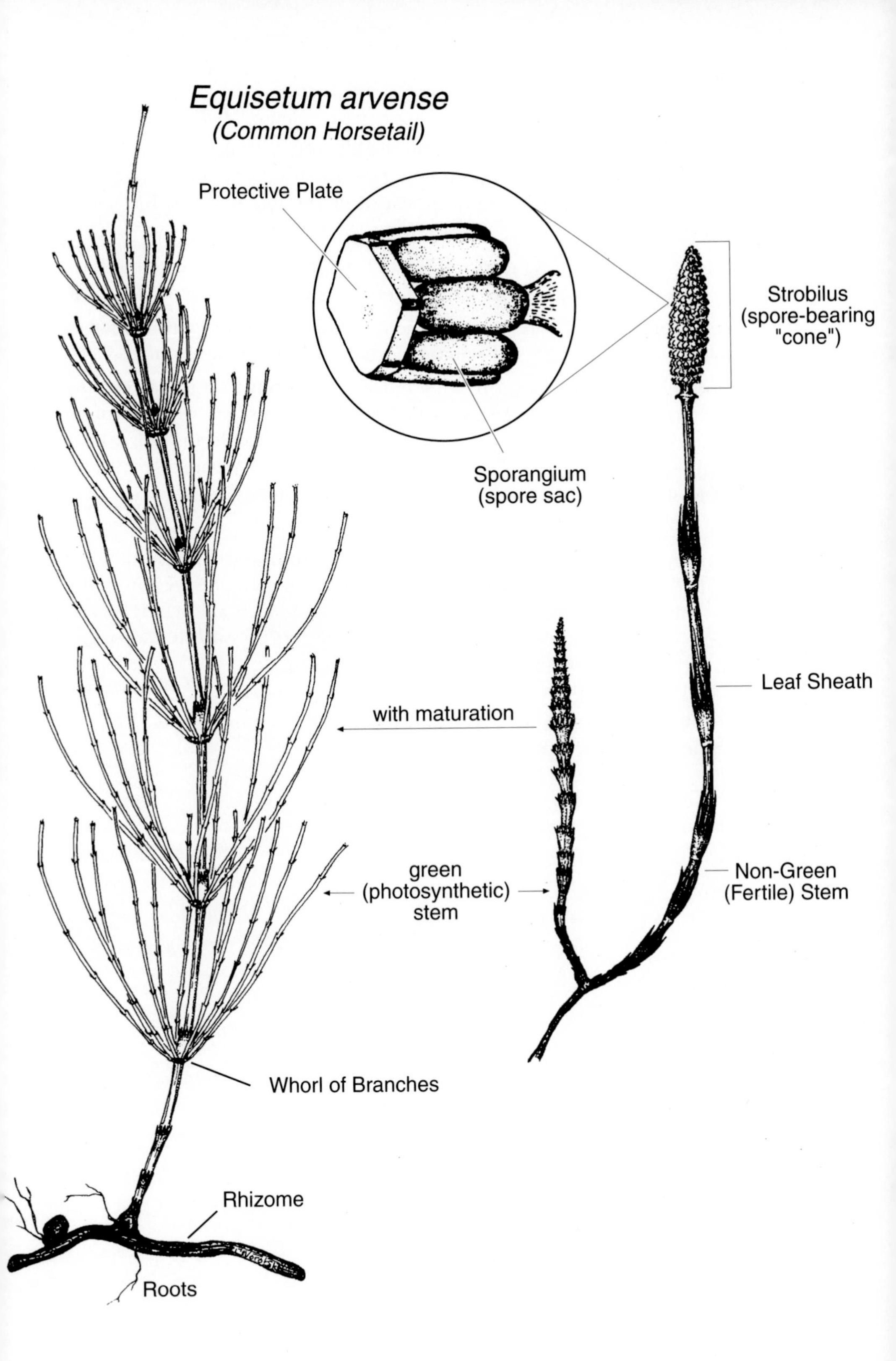
Equisetum arvense
(Common Horsetail)
Protective Plate
Strobilus
(spore-bearing
"cone")
Sporangium
(spore sac)
Leaf Sheath
with maturation
green
(photosynthetic)
stem
Non-Green
(Fertile) Stem
Whorl of Branches
Rhizome
Roots

Common Horsetail *Equisetum arvense*

The stems of common horsetail are dimorphic. Fertile stems are pale brownish, nonbranched, and usually about 6 inches (15 cm) high. Sterile stems are green, branched, much taller, and appear after the fertile stems have shed their spores.

Leaves: Sheath forming, with ten to twelve dark brown to black teeth on fertile stems, six to twelve green teeth on sterile stems.

Cones: Longer than wide, ½ to 1 inch (1 to 3 cm) long, dark brown to black, borne on stalks of equal or greater length at the stem tip.

Ecology: The most common horsetail in the region; ranges from lowlands, where it is often a weed, to high elevations, where it is usually found in seepage areas; it can grow in partial shade or full sun, in either dry or wet sites. The fact that the plants are extremely variable is probably an adaptive response to the varying environments.

Several horsetail species grow in the Pacific Northwest but primarily at low elevations, predominantly in wet places such as along streams or in floodplains. Of these, **common scouring-rush** *(E. hyemale)* is the best known. It is nondimorphic, with unbranched, evergreen fertile stems that are stiffly erect and often more than a meter tall. It occurs at low to mid-elevations and ranges from northern British Columbia through Washington.

A similar but smaller horsetail is **northern scouring-rush** (*E. variegatum),* which occurs more or less throughout the range, extending upward into the alpine zone in northern British Columbia.

Common Horsetail *Equisetum arvense*

Northern Scouring-Rush
Equisetum variegatum –George W. Douglas photo

Common Scouring-Rush
Equisetum hyemale –George W. Douglas photo

Common Scouring-Rush *Equisetum hyemale*

GRAPE FERNS (GRAPE FERN FAMILY) Ophioglossaceae

Grape ferns are unusual, attractive plants. The stems are short and completely buried beneath the surface of the ground. Each plant has a single, complex leaf consisting of a stalk with fertile and sterile segments. The fertile segment appears to be a continuation of the stalk and bears many large, grapelike spore sacs, the whole segment resembling a cluster of grapes. The sterile segment is variously divided, according to species, often parsleylike, branching from the leaf stalk at the base of the fertile segment.

There are several species of grape ferns, but most of them are rare and seldom encountered. The most common species in the Pacific Northwest is **leathery grape fern** *(Botrichium multifidum),* named for its leathery, evergreen sterile leaf segment. This species occurs sporadically throughout the region, from lowlands to well up in the mountains, mainly in open forests.

Leathery Grape Fern *Botrichium multifidum*

Leathery Grape Fern *Botrichium multifidum*

FERNS (FERN FAMILY) Polypodiaceae

Ferns constitute a highly diverse group of plants that taxonomists have separated into as many as five families. For convenience and simplicity, however, the traditional grouping of ferns into one family, Polypodiaceae, is followed here.

The northwest region of North America hosts a rich assortment of fern species, which frequently form dense populations in moist forest habitats. Most ferns have short, poorly developed underground stems called rhizomes; in some varieties, the rhizomes spread horizontally underground, from which new fronds arise. Aboveground, the fern plant consists of one or more leaves, or fronds. Fronds have a stipe (stalk), which is sometimes very short and is often covered with scales, and a blade. The blade is once to several times pinnate, that is, with pinnae or leaf segments arranged on opposite sides along the frond axis. Sword fern, for example, is once pinnate, and lady fern is two or three times pinnate. In the bud stages, the fronds are coiled, forming the well-known and edible fiddlehead.

Spore sacs cluster together in an area called a sorus (plural, sori) on the lower surface or along the margin of the frond. In most ferns, the sorus has a membranous shield or covering called an indusium, which varies in shape according to species. Each spore sac has a band of thick-walled cells that respond to water pressures, causing the spore sac to open and catapult its spores into the windstream, thus aiding dispersal.

The common ferns or fern groups of the region can be identified by using the key located in the back of the book; many of the technical terms are illustrated on the fern drawings.

Fern Leaf (Frond) Shape

Diagramatic

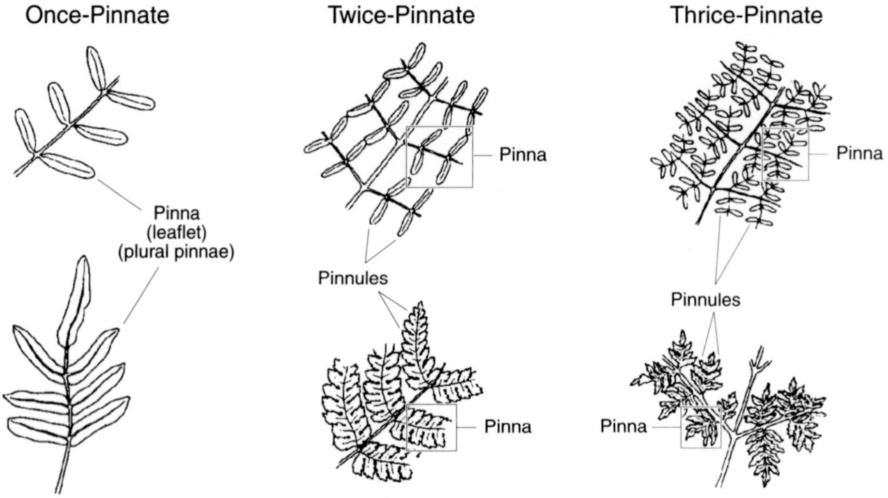

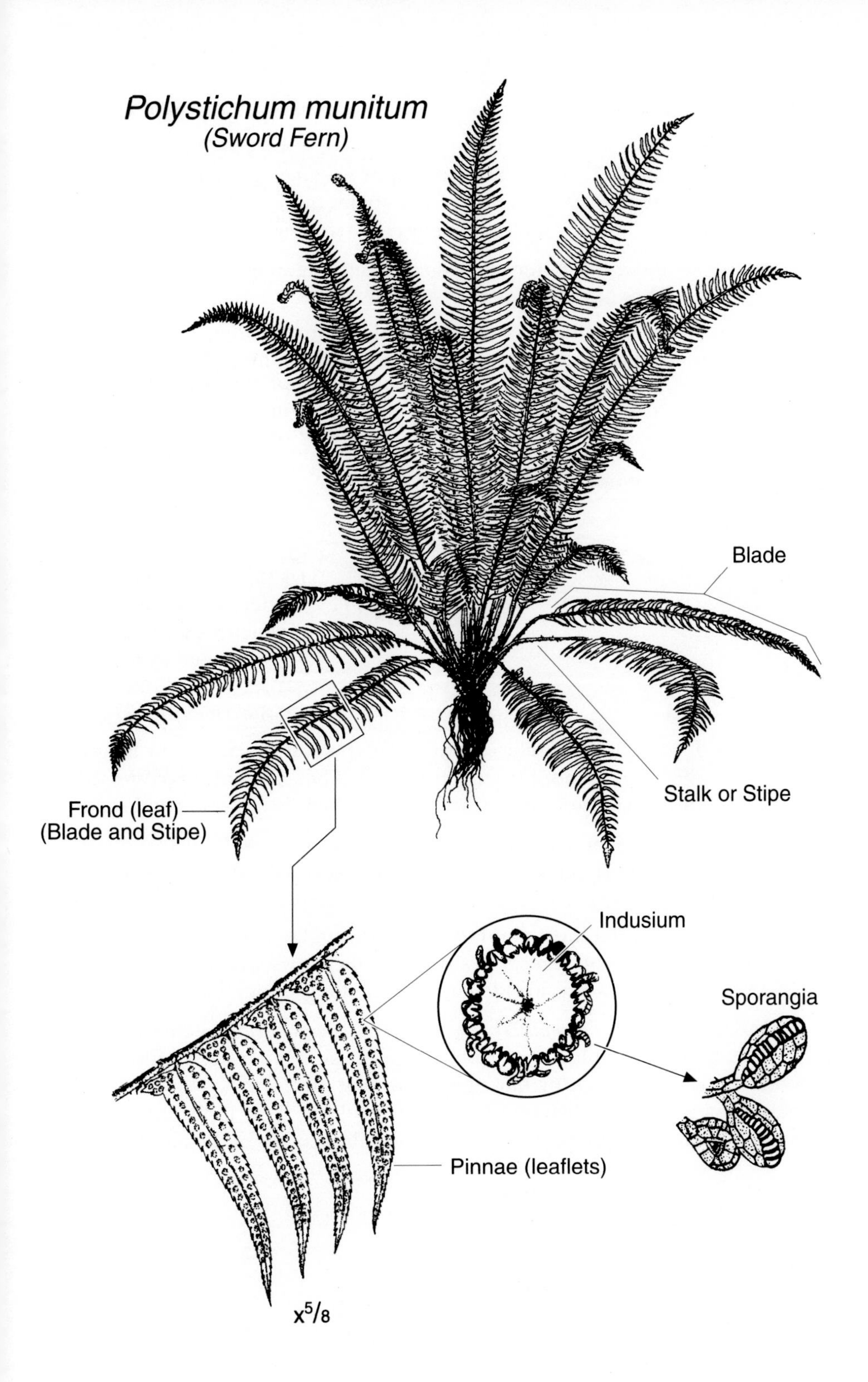
Polystichum munitum
(Sword Fern)
Blade
Frond (leaf)
(Blade and Stipe)
Stalk or Stipe
Indusium
Sporangia
Pinnae (leaflets)
x5/8

Maidenhair Fern

Adiantum aleuticum (pedatum)

Maidenhair ferns have fronds that arise singly or a few together from underground rhizomes. The leaf blades are kidney shaped, palmately divided, and oriented parallel to the ground at right angles to the stipe, which is purplish black, shiny, wiry, and up to 20 inches (50 cm) long.

Sori: Borne along the outer margins of frond segments and covered by reflexed lobes of the segments.

Ecology: This fern inhabits moist soils along streamsides, in seepages, and on wet rock outcrops or cliffs; shade tolerant and grows throughout the region, from lowlands to the alpine zone.

The maidenhair fern is unusual in appearance among ferns and is very attractive. It is especially abundant on rock faces along mountain streams where cascades envelope the ferns in mist.

Lady-Fern

Athyrium filix-femina

Lady-fern fronds occur in clusters of a few to several, varying in length from 6 inches (30 cm) to more than 3 feet (1 m). The leaf blades are broadly elliptical in outline, two to four times pinnate; the stipes are short and scaly.

Sori: Borne alongside conspicuous veins on the underside of frond segments; indusia flaplike and longer than wide, folded over the spore sacs from one side.

Ecology: One of the most common ferns throughout the Pacific Northwest, from lowland forests to subalpine meadows; often forms dense populations in moist, wooded areas and along stream banks.

Alpine lady-fern *(A. distentifolium),* a related species, is also common in the region. It grows mainly on wet, rocky slopes in subalpine and alpine areas and differs from lady-fern by having pinnae that are further separated (as suggested by the name "distentifolium") and by lacking an indusium. Also, the fronds are more tightly clumped and generally smaller in the alpine lady-fern.

Maidenhair Fern *Adiantum aleuticum*

Lady-Fern *Athyrium filix-femina*

Alpine Lady-Fern *Athyrium distentifolium*

Deer Fern *Blechnum spicant*

Deer fern fronds are once pinnate, clumped, and dimorphic (have two forms). The fertile fronds stand erect, 1 to 3 feet (3 to 9 dm) tall, and arise from the center of the clump. The sterile fronds spread horizontally. Pinnae of fertile fronds are linear, while those of sterile fronds are wider, rounded at the tip, and shiny.

Sori: Form a continuous line between the vein and outer margins of fertile frond pinnae; indusia thin and papery.

Ecology: Distributed throughout the region, preferring moist conifer forests and ranging from lowlands to high elevations in the mountains.

The attractive deer fern is distinctive because of its dimorphic fronds. Even in the absence of fertile fronds, it can easily be distinguished from other ferns that have once-pinnate fronds by its shiny, nontoothed, rounded pinnae.

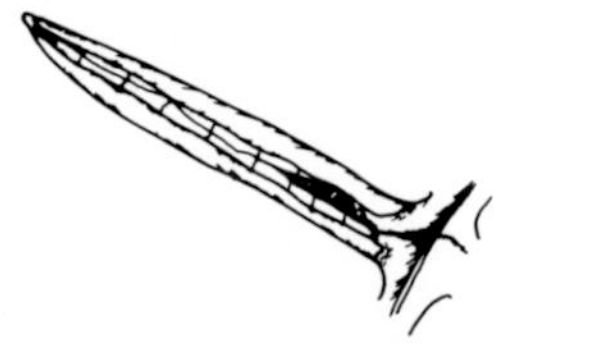

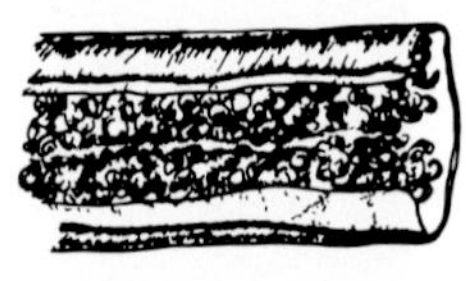

Parsley Fern *Cryptogramma acrostichoides (crispa)*

Parsley fern is distinctive, with evergreen, dimorphic fronds that form a dense but rather small clump. The fertile fronds consist of a relatively long stipe and a shorter blade divided into numerous linear segments. The sterile fronds are shorter and have parsleylike blades.

Sori: Arranged in two lines along the narrow frond segments, one on each side of the midvein; the margins of the segments are folded over the sori, forming one continuous indusium.

Ecology: A common fern on rocky slopes from lowlands to the alpine zone throughout the Pacific Northwest.

Indian's-dream fern *(Aspidotis densa)* resembles parsley fern, differing in that the fertile and sterile fronds are less distinct and the stipes are dark and shiny. It is much less common than parsley fern and has a more restricted distribution, ranging from southern British Columbia through Washington.

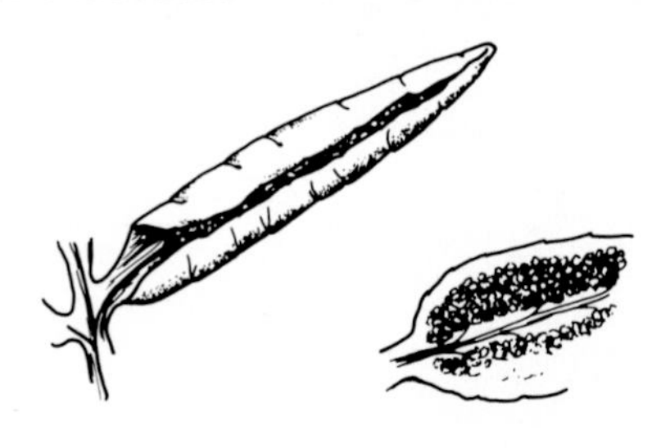

Deer Fern *Blechnum spicant*

Parsley Fern *Cryptogramma acrostichoides*

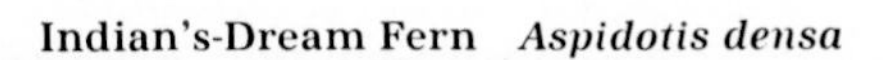

Indian's-Dream Fern *Aspidotis densa*

Lace Fern *Cheilanthes gracillima*

Lace fern is a low, spreading fern, 6 to 10 inches (15 to 25 cm) high, with numerous, densely clustered, evergreen fronds. The stipes are dark brown, almost as long as the blades. The blades are narrow and twice pinnate, with pale brown "wool" covering the lower surface.

Sori: Borne along the margins of the frond segments, obscured by "wool"; margins of frond segments fold inward, partially covering the sori and serving as a continuous indusium.

Ecology: Fairly common in southern British Columbia and Washington, growing in exposed rocky sites, especially along cliffs and in rock crevices on high mountain ridges.

Fragile Fern *Cystopteris fragilis*

Fragile fern produces loose clumps of a few to several small fronds, 4 to 12 inches (10 to 30 cm) long. The blades are two to three times pinnate, lance shaped in outline, and longer than the straw-colored stipes.

Sori: Borne along small veins of the frond segments; indusia thin, forming a pocket or pouch (*cystis* means "bladder") containing the spore sacs.

Ecology: Probably the most common Northwest fern of open areas, usually in rocky places; found throughout the region, from lowlands to alpine ridges, and although it prefers moist, open sites, it is also drought and shade tolerant.

Woodsia *Woodsia scopulina*

The fronds of woodsia are small, up to 8 inches (20 cm) long, tufted, and interspersed with persistent, older stipes. The blades are twice pinnate and longer than the brownish stipes. Whitish hairs cover the undersurface of the blades.

Sori: Rather large, occurring near the margin of the blade segments; indusia divided and starlike.

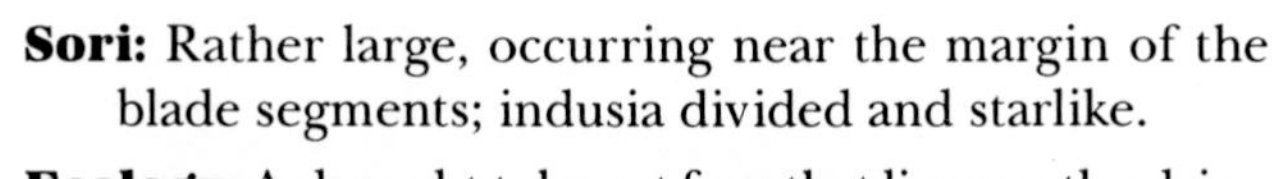

Ecology: A drought-tolerant fern that lives on the drier, eastern slopes of the Washington Cascades and western mountains of British Columbia, where it grows on rocky slopes and ledges.

Woodsia oregana, common in Washington and southern British Columbia, has a similar habitat.

Lace Fern *Cheilanthes gracillima*

Fragile Fern *Cystopteris fragilis*

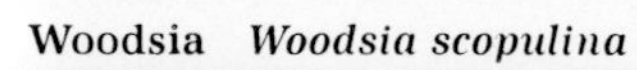

Woodsia *Woodsia scopulina*

Wood Fern

Dryopteris expansa (austriaca)

Wood fern fronds are three times pinnate and occur in clusters, from a few to several. The blades are triangular, 8 inches to 3 feet (2 to 9 dm) long, and may be as long or longer than they are broad. Brownish scales cover the stipes, which are shorter than the blades.

Sori: Borne at the ends of short veins that branch from the midvein of the frond segments; indusia shaped like a shield or a horseshoe.

Ecology: One of the most common forest ferns; distributed throughout the region, extending from lowlands to well up in the mountains; thrives in moist soils rich in organic material; usually grows on decaying stumps or logs.

Other species of *Dryopteris* in the region have narrower, often elliptical fronds that resemble those of lady fern but have horseshoe-shaped indusia rather than the elongate, flaplike indusia of lady-fern.

Oak Fern

Gymnocarpium dryopteris

Oak fern fronds are borne singly from underground rhizomes, are oriented horizontally, and stand 6 to 16 inches (15 to 40 cm) tall. The blades have three major stalked segments, pinnae, each of which is twice pinnate. The stipes are dull yellowish brown and are about as long as the triangular blade.

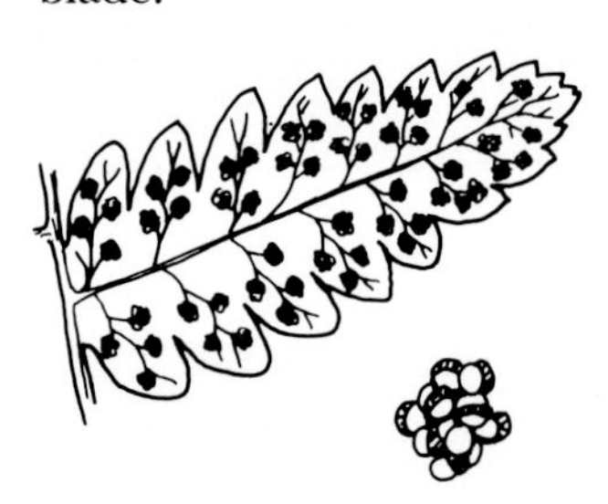

Sori: Small and numerous along the veins of the frond segments; indusia absent.

Ecology: Distributed throughout the Pacific Northwest in conifer forests at all elevations.

The oak fern resembles the wood fern, as suggested by the shared Latin name, *dryopteris,* but can easily be distinguished by its solitary, horizontally oriented frond with three primary segments, and by the absence of an indusium. Also, oak fern spreads along the moist forest floor rather than growing on decaying wood.

Goldback fern *(Pityrogramma triangularis)* superficially resembles the oak fern and wood fern but is loosely clumped, has black, shiny stipes, and is covered on the lower surface of the frond blade with golden yellow powder. The goldback fern is restricted to rocky exposed sites along the coast and low elevations in the mountains.

Wood Fern *Dryopteris expansa*

Oak Fern *Gymnocarpium dryopteris*

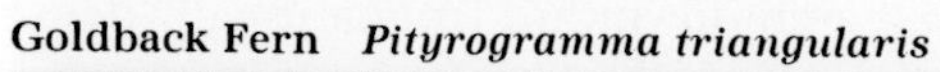

Goldback Fern *Pityrogramma triangularis*

Licorice Fern *Polypodium glycyrrhiza*

Licorice fern gets its name from the strong licorice-like taste of the thick, scaly rhizomes. As with other members of the genus, the branched rhizomes fancifully resemble feet (*Polypodium* means "many feet"). This fern spreads by branching rhizomes that creep over tree and rock substrates and when exposed are conspicuous with a dense covering of pale brown scales. The 8-to-16-inch (20 to 40 cm) fronds are once pinnate, with the pinnae nonstalked and widest at the base. The blades are usually slightly longer than the stipes.

Sori: Large and conspicuous, in two rows–one on each side of the pinnae midvein; indusia absent.

Ecology: Licorice fern grows primarily on trees, particularly big-leaf maple, and on rocks, usually associated with mosses; its range extends inward and upward from the coast to low elevations in the mountains.

Cliff or **western polypody** *(P. hesperium)* resembles licorice fern but has rounded, rather than pointed, pinnae and smaller, more leathery fronds. This species grows along rock ledges and in rock crevices, from lowlands to mid-elevations in the mountains.

Sword Fern *Polystichum munitum*

Because of its abundance and wide distribution, sword fern is probably the best-known fern in the Pacific Northwest. Its broad ecological tolerance allows it to grow in dry as well as moist sites and in shade as well as in full sunlight. Sword fern is highly variable because of a combination of genetic, environmental, and growth factors. The fronds are often more than 3 feet (1 m) long, are once pinnate, and are narrowly lance shaped in outline. They project outward and arch downward from a usually dense cluster. The pinnae are sharp pointed and toothed, with a prominent earlike lobe adjacent to the distinct stalk. Pale brown scales cover the stipe.

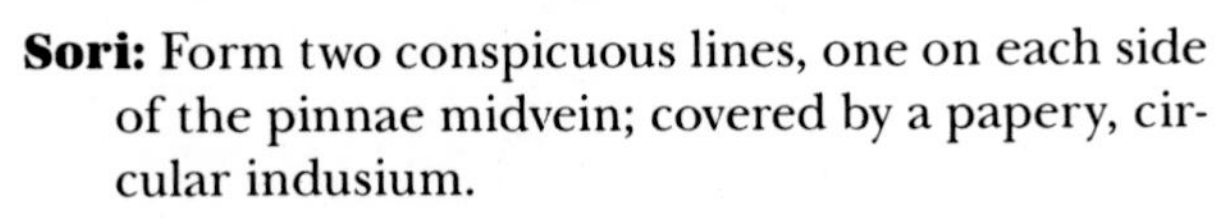

Sori: Form two conspicuous lines, one on each side of the pinnae midvein; covered by a papery, circular indusium.

Ecology: Abundant in lowland to mid-elevation forests and forest openings throughout the range.

In the juvenile state, sword fern closely resembles **holly fern** *(P. lonchitis),* which has softer, narrower fronds and shorter, more prominently toothed pinnae. Holly fern occupies mainly rocky habitats, from mid-elevations to the alpine zone.

Licorice Fern *Polypodium glycyrrhiza*

Cliff Polypody *Polypodium hesperium*

Sword Fern *Polystichum munitum*

Shasta Fern

Polystichum lemmonii (mohroides)

Shasta fern has soft, narrow fronds about 16 inches (40 cm) long, usually densely clustered. The pinnae are crowded and overlap and have the unusual characteristic of being deeply divided and twice pinnate, especially toward the base of the blade. The stipes are densely and conspicuously covered with brown scales.

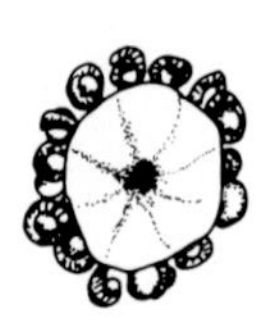

Sori: Numerous and irregularly aligned along the pinnae midvein; covered by circular indusia.

Ecology: An attractive but uncommon fern that rarely extends northward beyond Washington; found in exposed, rocky places in the mountains, often associated with serpentine (magnesium rich, calcium poor) rock.

A species somewhat similar in appearance and habitat preference is **Kruckeberg's fern** *(P. kruckebergii).* However, its pinnae are less deeply lobed and have spine-tipped teeth. It is distributed along the coastal mountains from Washington into southeastern Alaska. Another species with less divided pinnae and spine-tipped teeth is **rock sword fern** *(P. scopulinum),* which grows in rocky places, as the common name suggests, and is distributed along the eastern side of the Cascade Range in Washington and southern British Columbia.

Bracken Fern

Pteridium aquilinum

Bracken fern fronds are large and coarse, up to 10 feet (3 m) tall, and grow singly from spreading underground rhizomes. The blades are three or four times pinnate, with major segments stalked; they are oriented more or less parallel to the ground and perpendicular to the stipe, which is green to straw colored and of nearly equal length. Dense, fine hairs cover the undersurface of the blade segments.

Sori: Numerous and small, aligned along the margins of the frond segments, the margins then folded over the sori to provide a continuous, protective indusium.

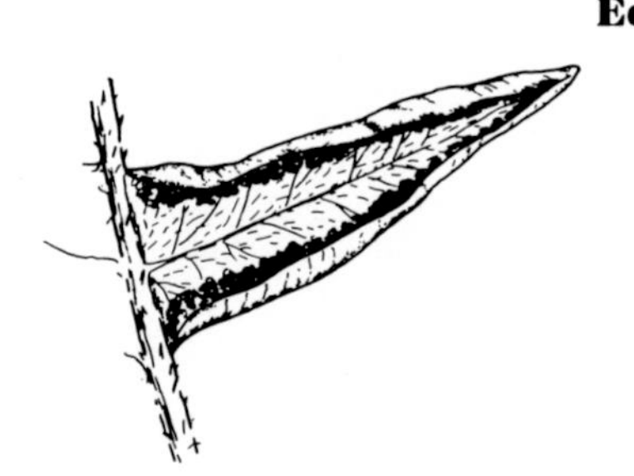

Ecology: A weedy species, most frequently observed in open, disturbed sites, such as recently logged areas; its broad ecological tolerance enables it to grow in a variety of natural habitats, including dense and often dry forests; distributed throughout the Pacific Northwest, from lowlands to mid-elevations in the mountains.

Shasta Fern *Polystichum lemmonii*

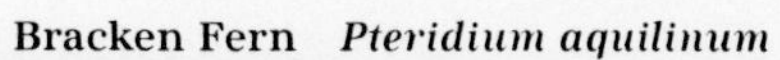

Bracken Fern *Pteridium aquilinum*

Yellow Aster Butte, North Cascades

TREES

"A woody perennial plant having a single main axis or stem (trunk), commonly exceeding 10 feet in height." Webster's definition of *tree* is simplistic, but generally applicable. A few large shrubs, such as vine maple, Scouler willow, and Sitka alder, frequently are grouped with trees, even though they typically have multiple stems. We have followed this convenient grouping here.

The key located in the back of the book is suggested as an aid to the identification of the trees in the region.

Conifers

Perhaps more than anything else, the magnificent conifers distinguish the Northwest forest, which extends from the Pacific coast to the alpine treeline. Two categories of conifers dominate these forests: **climax** species and **seral** species. Climax species represent the final stage of succession; they are sufficiently shade tolerant that other species will not replace them, and they can perpetuate themselves indefinitely. Seral species are trees that invade open spaces but ultimately yield to the more shade-tolerant climax species. Zonation of the conifer forests is discussed in the Introduction.

CYPRESS FAMILY — Cupressaceae

The cypress family comprises evergreen trees and shrubs (junipers) with small, opposite, scalelike leaves. In some junipers and juvenile cedar shoots the leaves are sometimes whorled and more or less needlelike. Seeds are borne in woody or berrylike (junipers) cones.

Alaska Cedar — *Chamaecyparis nootkatensis*

Alaska cedar is a beautiful and commercially valuable tree with tough, straight-grained wood that is highly resistant to decay. In outline, the tree has a distinctive profile, with a characteristic slender crown and drooping branches. The bark is thin and grayish.

Leaves: Scalelike, opposite in four rows, with a somewhat conspicuous longitudinal ridge (keel) on the back, and a prickly tip.

Cones: Woody, round, less than ½ inch (12 mm) thick; scales attached in the middle (shield shaped) with a central point.

Ecology: A tree of moist, cool habitats at middle to high elevations; usually in stands with Pacific silver fir and mountain hemlock on the western slopes of the mountain ranges, often extending up to the treeline. In northern British Columbia and southeastern Alaska, it grows mainly as a small tree in bogs.

Alaska cedar can easily be distinguished from western red cedar by its round cones, whitish bark, and slender, drooping branches.

Alaska Cedar crowns
Chamaecyparis nootkatensis

Alaska Cedar trunk
Chamaecyparis nootkatensis

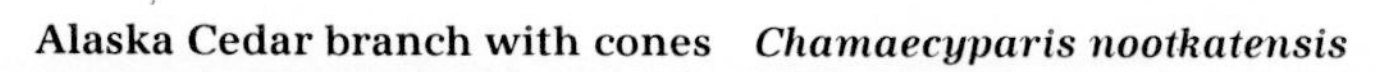

Alaska Cedar branch with cones *Chamaecyparis nootkatensis*

Western Red Cedar *Thuja plicata*

Western red cedars can grow to be very large trees, often more than 10 feet (3 m) in diameter and 200 feet (70 m) tall. The crown is cone shaped, with spraylike branches spreading down and outward. The bark is thin and gray to reddish. The trunk is typically buttressed.

Western red cedar was probably the most important tree to the Northwest coastal Indians. The bark shreds easily and was used for making baskets and clothing. The wood is rot resistant, splits easily, and was made into a multitude of items, from dishes to canoes and from ceremonial masks to totem poles. The trees were equally important to European colonists. Today, old-growth cedars are becoming scarce and extremely valuable commercially, particularly in Washington.

Leaves: Scalelike, lying flat against the twigs, opposite in four rows, rounded on the back, and sharp pointed.

Cones: Woody, egg shaped or cylindrical, less than ½ inch (12 mm) thick and about ½ inch (15 cm) long; scales egg shaped and attached at the base.

Ecology: Grows best at low elevations in cool, moist habitats on the western slopes of the mountain ranges in the region; along with western hemlock, the species is often a climax dominant, extending upward along major drainages to 4,000 to 5,000 feet (1,219 to 1,524 m).

Mountain Juniper *Juniperus communis*

Mountain Juniper is a low, spreading, matted shrub, seldom more than 3 feet (1 m) tall.

Leaves: About ½ inch (1 to 2 cm) long, more or less needlelike with a spiny tip, growing in whorls of three per node.

Cones: Berrylike, round but roughened, about ¼ inch (6 to 8 mm) thick.

Ecology: A highly variable species, both in growth form and habitat; the most common variety, distributed throughout the Pacific Northwest, is a low shrub that grows on open, rocky ridges from mid-elevations upward into the alpine zone.

Western Red Cedar trunk
Thuja plicata

Mountain Juniper branch and berries
Juniperus communis

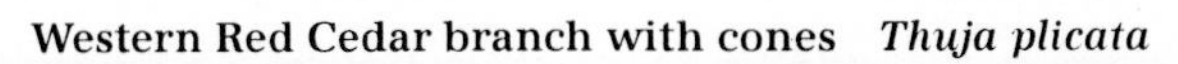

Western Red Cedar branch with cones *Thuja plicata*

Pine Family

Pinaceae

The pine family comprises trees with needlelike leaves (usually evergreen) that are borne either singly and arranged in a spiral on the twigs, or in clusters of two or more. Seeds are produced in woody cones, with spirally arranged cone scales.

Pacific Silver Fir

Abies amabilis

Pacific silver fir is a large tree, up to 200 feet (60 m) tall and 4 feet (1.2 m) in diameter. Its bark is thin, smooth, and grayish, becoming fissured and often covered with moss on old trees. The crowns are narrowly conic and symmetrical. This species is aptly named—*amabilis* means "beautiful," and *silver* relates to the color on the underside of the leaves, which contrasts sharply with the bright green upper surface.

Leaves: Shiny green on top, whitish beneath, slightly notched at the tip, usually less than 1 inch (2.5 cm) long; longer needles spread to the side, making the branches appear flattened and spraylike; shorter needles are concentrated along the top of the twigs and project forward, hiding the twigs.

Cones: Erect in the treetops, barrel shaped, purplish, 3 to 6 inches (8 to 15 cm) long; scales deciduous, falling away from the persistent cone axis.

Ecology: Tolerates both shade and deep snow, so it has become a climax dominant species over a broad elevational range on the western mountain slopes of Washington and southern British Columbia; because of its dominance, some plant ecologists refer to the region as the Pacific silver fir zone; also invades subalpine meadows and is a frequent component of tree clumps.

Pacific silver fir can be distinguished from other true firs (*Abies* species) by the characteristics and orientation of the needles. Stands of Pacific silver fir usually include western or mountain hemlock.

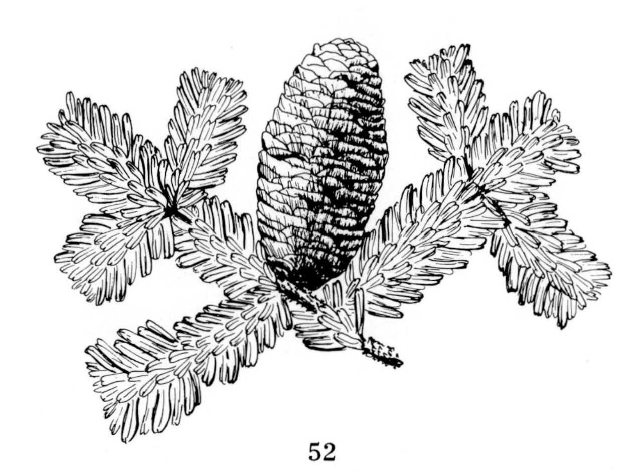

Pacific Silver Fir trunk
Abies amabilis

Pacific Silver Fir silhouette
Abies amabilis

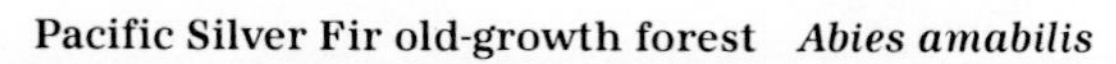

Pacific Silver Fir old-growth forest *Abies amabilis*

Grand Fir *Abies grandis*

Grand fir is the largest true fir, up to 250 feet (75 m) tall and 5 feet (1.5 m) in diameter. The bark is dark gray, thin, and smooth in young trees, thicker and grooved with age. The crowns are cone shaped.

Leaves: Shiny green on top, whitish beneath, notched at the tip, usually more than 1 inch (2.5 cm) long, spread sideways, making the branches appear flattened and spraylike.

Cones: Erect in the treetops, cylindrical, greenish, 3 to 5 inches (8 to 12 cm) long; cone scales deciduous.

Ecology: Primarily a seral species, sporadically distributed at low to mid-elevations in the mountains of Washington and southern British Columbia; most common on the eastern side of the Cascade crest, where it may be locally dominant.

Grand fir boughs resemble the silver fir because the needles are silver on the underside and bright green on top. But the needles of grand fir all project to the side, which, unlike silver fir, make the branch and twigs easily visible from above.

Grand Fir trunk ***Abies grandis***

Noble Fir *Abies procera*

Noble fir is a rather large tree, up to 200 feet (60 m) tall and 3 feet (1 m) in diameter, with a narrow, symmetrical crown. The bark is thin and grayish.

Leaves: Pale green to whitish on both surfaces, rounded at the tip, usually more than 1 inch (2.5 cm) long, curving upward on the twigs.

Cones: Erect in the treetops, barrel shaped, brownish, 4 to 7 inches (10 to 18 cm) long, with conspicuous extended bracts; scales deciduous.

Ecology: A common tree at middle to upper elevations in the continuous forest zone of the southern Cascades in Washington and Oregon.

When cones are not present, noble fir is difficult to distinguish from subalpine fir. However, the needles of noble fir are longer and thicker than subalpine fir's and are directed upward rather than outward in all directions. Also, noble fir is a larger tree and has a broader crown. Noble fir is an important lumber tree at middle to high elevations in the southern Cascades, but it is probably best known for its value in the Christmas tree market. The beautifully symmetrical crown makes it a favorite, and it is widely cultivated in tree farms.

Noble Fir *Abies procera*

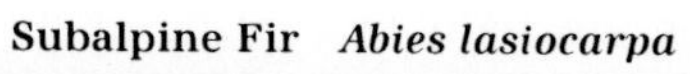

Subalpine Fir *Abies lasiocarpa*

Subalpine Fir *Abies lasiocarpa*

Subalpine fir is a small to medium-sized tree, with thin, blistered, grayish bark that becomes fissured as the tree ages. The crowns are distinctively narrow and spirelike.

Leaves: Pale green to whitish on both surfaces, rounded at the tip, usually less than 1 inch (2.5 cm) long, spreading more or less around the twig, giving the twigs a brushlike appearance.

Cones: Erect in the treetops, cylindrical, purplish, 2 to 4 inches (6 to 10 cm) long.

Ecology: A high-elevation mountain species, often growing in clumps in and around subalpine meadows and upward into the alpine zone as a krummholz species; often a climax dominant in the upper continuous forests on the east-side mountain ranges of Washington and southern British Columbia; distributed throughout the region but only sporadically in northern British Columbia and southeastern Alaska.

Subalpine fir grows in all the mountain ranges of western North America, from Alaska southward into the California Sierra Nevada and the Rockies of southern Arizona and New Mexico. As the common name suggests, subalpine fir is mainly a high-elevation species, but it frequently extends downward to mid-elevations in cold-air drainages. In glacial valleys in northern British Columbia and southeastern Alaska, it persists primarily in krummholz form, reproducing asexually by rooting of spreading branches (layering). Over its broad geographical and ecological range it is extremely variable, and a number of varieties or subspecies are recognized.

Subalpine Fir crowns *Abies lasiocarpa*

Subalpine Fir trunk *Abies lasiocarpa*

Subalpine Fir foliage *Abies lasiocarpa*

Subalpine Larch *Larix lyallii*

Subalpine larch is a small tree, up to 50 feet (15 m) tall, with thin, furrowed bark and a narrow, ragged crown. The twigs are densely hairy and roughened by spur shoots that bear dense clusters of leaves. This species provides a colorful autumn display, when its golden orange leaves contrast with the dark green mountain slopes.

Leaves: Borne in clusters of twenty to thirty on spur shoots, soft, deciduous.

Cones: Small, about 1 inch (2.5 to 3.5 cm) long, with extended bracts, borne throughout the tree.

Ecology: A small tree that grows along the crest of the Cascades of Washington and southern British Columbia, particularly on north-facing slopes on the eastern side, extending upward from the treeline into the alpine zone.

Subalpine larch commonly grows in association with krummholz forms of mountain hemlock, Engelmann spruce, and subalpine fir but retains its short, upright form.

Western Larch *Larix occidentalis*

Western larch is a large tree, often more than 200 feet (60 m) tall and 3 feet (1 m) in diameter, with a narrow, ragged crown. Its bark is thick and deeply furrowed, and the twigs are hairless to lightly hairy, roughened by spur shoots that bear the leaves.

Leaves: Borne in clusters of twenty-five to forty on spur shoots, soft, deciduous.

Cones: Small, about 1 inch (2.5 to 3.5 cm) long, with extended bracts, borne throughout the tree.

Ecology: A middle to high-elevation tree on the eastern side of the Washington Cascades, extending northward into southern British Columbia; typically grows in association with other conifers, which might include Douglas fir, grand fir, white pine, Engelmann spruce, and lodgepole pine.

Western larch, with its thick, protective bark, is highly resistant to fire. The wood is dense and durable, making it a valuable lumber tree. In all seasons it stands out from the other trees in the forest: in winter it has no leaves; in autumn the leaves are golden orange, and in spring and summer the leaves are soft, light green.

Subalpine Larch with early growth of leaves and cones *Larix lyallii*

Western Larch branch with cones *Larix occidentalis*

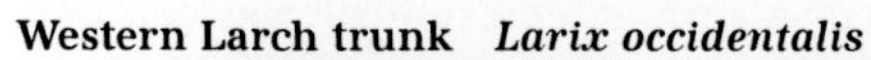

Western Larch trunk *Larix occidentalis*

Engelmann Spruce *Picea engelmannii*

Engelmann spruce is a large tree, up to 200 feet (60 m) tall and 3 feet (1 m) wide, with a narrow, ragged crown. The bark is thin, gray, and scaly. As in all spruces, the twigs are rough because of persistent leaf bases.

Leaves: Quadrangular, stiff and sharp pointed, whitish bands similar on all four sides.

Cones: Borne primarily in the treetops; 1 to 2½ inches (3 to 6 cm) long; scales more or less diamond shaped, with jagged tips.

Ecology: A climax dominant on the cool, moist slopes of the eastern Cascades in Washington and adjacent British Columbia, extending up to the alpine zone; usually grows with subalpine fir.

Engelmann spruce is widely distributed in the mountains of western North America and is extremely variable, taking on different forms depending on where it is growing. To add to the taxonomic complexity of the species, it hybridizes with other spruces. In British Columbia, a broad hybrid zone exists where the Engelmann and white spruce ranges overlap.

Sitka Spruce *Picea sitchensis*

Sitka spruce can become exceedingly large, more than 250 feet (75 m) tall and 12 feet (3.5 m) in diameter. The trees have a narrow, ragged crown and thin, gray, scaly bark. Persistent leaf bases make the twigs rough.

Leaves: Four-sided but flattened and more or less triangular in cross section, stiff and spiny, white bands much broader and more conspicuous on the upper side of the needles.

Cones: Borne primarily in the treetops, 2 to 4 inches (5 to 10 cm) long; scales more or less diamond shaped, with jagged edges.

Ecology: A species of lower, western mountain slopes, extending inward along major river drainages in British Columbia and southeastern Alaska; primarily a seral species in the western hemlock zone.

This is the big tree in the Olympic "rain forests," along the coast of southern British Columbia and Vancouver Island, and on the Queen Charlotte Islands. Along river drainages in British Columbia, it frequently comes in contact with Engelmann spruce, resulting in hybrid swarms.

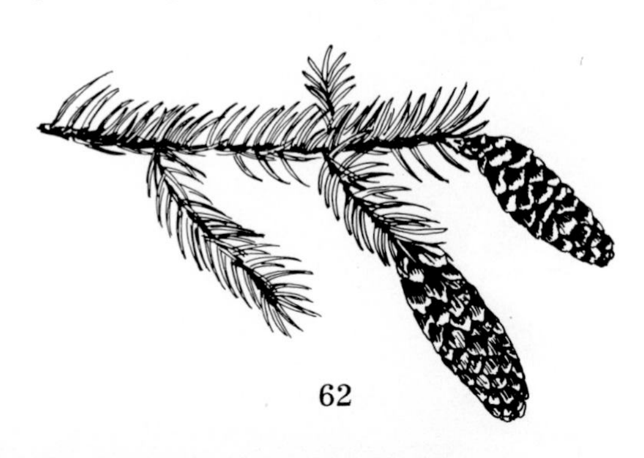

Engelmann Spruce trunk *Picea engelmannii*

Engelmann Spruce cones *Picea engelmannii*

Sitka Spruce *Picea sitchensis*

White Spruce *Picea glauca*

White spruce is a medium-sized tree, up to 80 feet (25 m) tall and 2 feet (60 cm) in diameter, with a narrow, ragged crown and thin, gray, scaly bark. Persistent leaf bases make the twigs rough.

Leaves: Quadrangular, stiff and sharp pointed, whitish bands similar on all four sides.

Cones: Borne primarily in the treetops, 2 to 3 inches (5 to 8 cm) long; scales fan shaped, broadest near the smooth tip.

Ecology: The major dominant species in boreal forests of Canada and Alaska; in the Pacific Northwest, found primarily on the eastern side of the northern mountain ranges.

White spruce hybridizes with Engelmann spruce in southern British Columbia, resulting in extensive hybrid swarms. It also hybridizes with Sitka spruce along the major river drainages in northern British Columbia and southeastern Alaska.

Black spruce *(P. mariana)* also inhabits northwestern British Columbia and southeastern Alaska but is much more common in the Alaskan and Canadian interiors. Although black spruce resembles white spruce, it is smaller in every respect and is seldom more than 30 feet (9 m) tall. The cones too are smaller and rounder. Black spruce prefers boggy (muskeg) habitats, mosquito heaven. Along the muskeg margins, it hybridizes with white spruce.

White Spruce forest *Picea glauca*

White Spruce branch with cones *Picea glauca* –George W. Douglas photo

Black Spruce bog *Picea mariana*

Whitebark Pine *Pinus albicaulis*

Whitebark pine is a small tree, up to 60 feet (18 m) tall and 2 feet (60 cm) in diameter, usually freely branched, with a broad, spreading crown. The bark in young trees is smooth and whitish, becoming gray and scaly with age.

Leaves: Borne in clusters of five, rather stiff, yellow-green, 2 to 3 inches (5 to 8 cm) long.

Cones: Egg shaped; 1½ to 3 inches (4 to 8 cm) long, tending to remain unopened on the trees for several years; scales have a prominent terminal point.

Ecology: A species of the high mountains, growing near or above the treeline; ranges from southern British Columbia through Washington, primarily east of the Cascade crest.

Whitebark pine prefers rocky subalpine meadows, where it achieves maximum size. At higher elevations, above treeline, it becomes dwarfed and shrublike. On open ridges, where they are most frequently found, the trees become twisted as they are shaped by high winds and drifting snow, which gives them a distinctive grotesque beauty.

At the upper limit of the continuous forest, this species often mixes with western white pine, which also has needles in bundles of five. Although young trees of the two species may look similar, western white pine is distinguishable by its more flexible, bluish green needles. Whitebark pine resembles western white pine in being susceptible to white pine blister rust and usually succumbs to this disease at lower elevations.

Whitebark Pine trunk *Pinus albicaulis*

Whitebark Pine, alpine (krummholz) form *Pinus albicaulis*

Whitebark Pine meadow *Pinus albicaulis*

Lodgepole Pine *Pinus contorta*

A medium-sized tree, lodgepole pine grows to 100 feet (30 m) tall and 3 feet (1 m) in diameter, with a crown that is rounded in open habitats and cylindrical in forested areas. The bark is gray to reddish brown, scaly, and often pitchy.

Leaves: Borne in clusters of two or occasionally three; yellow-green, about 1 to 2½ inches (3 to 6 cm) long, rather stiff.

Cones: Usually bent and narrowly egg shaped, 1½ to 2½ inches (4 to 6 cm) long, remaining on the trees for several years, often unopened; a sharp prickle stands out from the back of the cone scales.

Ecology: Lives in a variety of habitats, from the coast to near treeline; distributed throughout the Pacific Northwest except in the St. Elias ranges of British Columbia and adjacent Alaska.

Lodgepole pine is a variable species with an extremely broad ecological tolerance and can grow in habitats unsuited to other conifers. For example, it tolerates the wet, acidic environment of bogs and the dry, toxic soils underlain by serpentine rock. It can also grow on thin, volcanic soils and is a seral species more or less throughout its range. In some trees, the cones remain closed until fire heats them, releasing the seeds. This is why burned areas are quickly repopulated by lodgepole pine. A vast expanse of lodgepole pine forest suggests a fire history.

Lodgepole Pine cones *Pinus contorta*

Lodgepole Pine trunk *Pinus contorta*

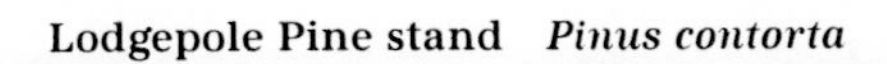

Lodgepole Pine stand *Pinus contorta*

Western White Pine *Pinus monticola*

Western white pine may grow to be very large, more than 250 feet (75 m) tall and 5 feet (1.5 m) in diameter. The crowns are cylindrical and often ragged. The bark is gray and thin, marked by randomly oriented grooves.

White pine is a valuable lumber tree that was once broadly distributed across the mountains of Washington and southern British Columbia. However, populations have been depleted by white pine blister rust; these days, trees rarely achieve their full growth potential before the rust kills them. White pine is a popular ornamental tree, but seldom escapes the rust. Efforts are under way in forest research laboratories to breed rust-resistant trees.

Leaves: Borne in clusters of five, blue-green, flexuous, 2½ to 4 inches (6 to 10 cm) long.

Cones: Usually curved, 6 to 10 inches (15 to 25 cm) long; cylindrical, mainly occurring near the treetops, not retained on the trees after maturity; scales typically resinous.

Ecology: An occasional tree in the Pacific Northwest, at middle to high elevations; most common in Douglas fir and Engelmann spruce forests on the eastern side of the Cascade crest and the Coast Mountains of southern British Columbia.

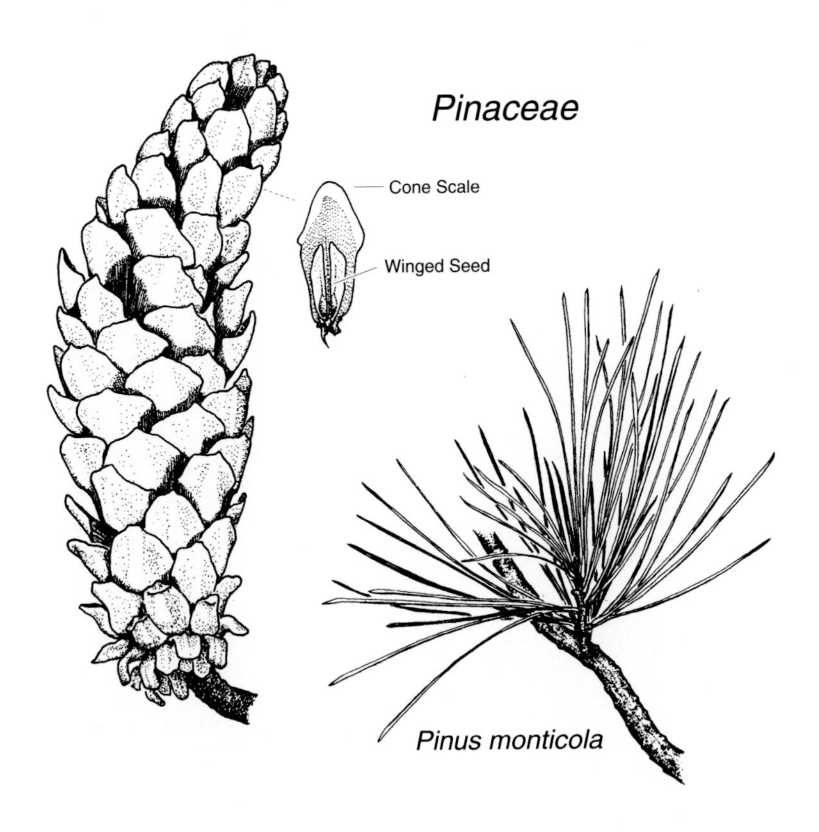

Western White Pine foliage with cones *Pinus monticola*

Western White Pine trunk
Pinus monticola

Western White Pine crown
Pinus monticola

Ponderosa Pine *Pinus ponderosa*

Ponderosa pine is a large tree, as the name suggests, exceeding 200 feet (60 m) tall and 4 feet (1.2 m) in diameter. The crown is round to broadly cylindrical. The bark is gray-brown, becoming reddish brown and deeply furrowed in larger trees, with jigsaw-puzzle-shaped plates freely sloughing off.

Ponderosa pine has an interesting relationship with Douglas fir. It is not shade tolerant, and Douglas fir tends to replace it in areas where soil moisture is not a limiting factor. Douglas fir, on the other hand, is less tolerant of fire. Therefore, many or most ponderosa pine forests are dependent on periodic fires for their persistence. Both species are valuable lumber trees.

Ponderosa pine is a favorite target of the pine-bark beetle, which routinely kills older trees, particularly those weakened from periodic drought. Few stands escape the ravages of this insect.

Leaves: Borne in clusters of three or occasionally two, dark green, 5 to 10 inches (12 to 25 cm) long.

Cones: Broadly egg shaped and symmetrical, 3 to 5 inches (8 to 12 cm) long; prominent prickles arm the back of the cone scales.

Ecology: The most drought-tolerant tree in the region, dominating the forest zone immediately above the sagebrush steppe on the eastern side of the Washington Cascades and adjacent British Columbia. It extends upward to high elevations on dry, rocky, south-facing slopes.

Ponderosa Pine stand *Pinus ponderosa*

Ponderosa Pine trunk *Pinus ponderosa*

Ponderosa Pine cones *Pinus ponderosa*

Western Hemlock *Tsuga heterophylla*

Western hemlock is a large tree, up to 175 feet (53 m) tall and 4 feet (1.2 m) in diameter. The crown is narrowly cone shaped, with a drooping tip, particularly in young trees. The bark is gray-brown, with shallow, scaly ridges. Short, persistent leaf stalks make the twigs rough.

This is probably the most shade-tolerant tree in the Pacific Northwest. It can reproduce in closed forests, where it will ultimately replace other tree species. At upper elevations, it in turn is replaced by mountain hemlock, which is more cold tolerant and capable of surviving deeper snowpack.

Western hemlock is an attractive tree, particularly when young, with delicate, weeping branches and a drooping crown. It is valuable commercially and is the Washington state tree.

Leaves: Dark green on top and whitish beneath, variable lengths (as the Latin name *heterophylla* suggests), up to 1 inch (2.5 cm) long; they project outward on the sides of the twigs, making the branches appear flattened and spraylike.

Cones: Small, up to 1 inch (1.5 to 2.5 cm) long and nearly as wide.

Ecology: A climax dominant species across much of its range, from sea level to mid-elevations on the moist, western side of the mountain ranges; distributed from southeastern Alaska south through the Washington Cascades.

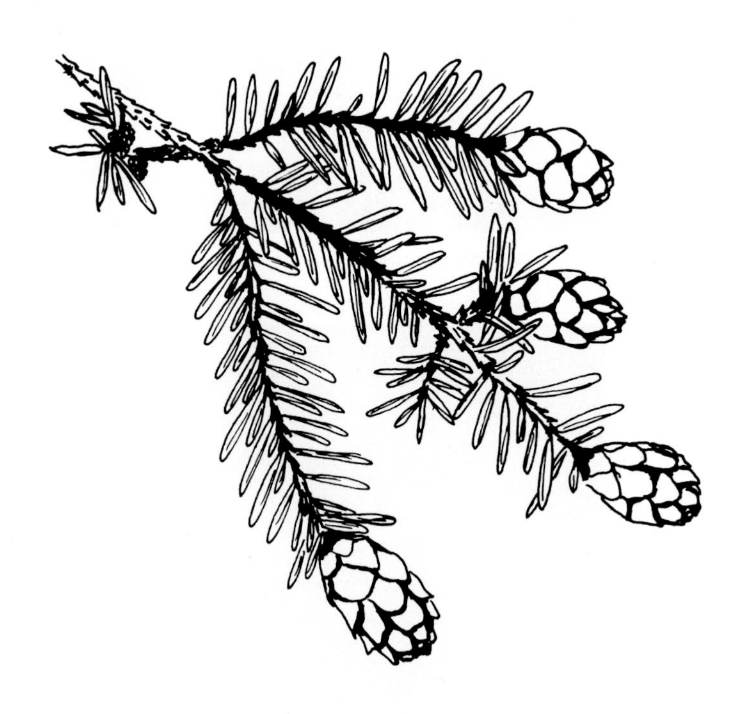

Western Hemlock foliage with cones *Tsuga heterophylla*

Western Hemlock trunk *Tsuga heterophylla*

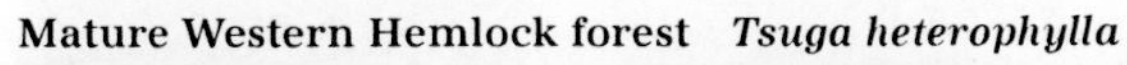

Mature Western Hemlock forest *Tsuga heterophylla*

Mountain Hemlock *Tsuga mertensiana*

Mountain Hemlock, a medium-sized tree, grows up to 100 feet (30 m) tall and 3 feet (1 m) in diameter, with a broadly cone-shaped crown. The bark is dark brown, with deep ridges, particularly in older trees. Persistent leaf bases make the twigs rough.

Leaves: Pale green to whitish on both sides, more or less the same length, about ¾ inch (2 cm) long, projecting outward in all directions on the twigs, brushlike.

Cones: Broadly cylindrical, 1 to 2 inches (2.5 to 4 cm) long.

Ecology: A climax dominant species on the cool, moist upper mountain slopes, extending upward to the alpine zone, where it grows in krummholz form; mainly restricted to the moist western side of the mountain ranges, from southeastern Alaska through Washington.

The ranges of mountain and western hemlock overlap in elevation and in latitude, and the two species resemble each other. However, mountain hemlock has larger cones, more rigid branches and a more upright crown, more uniform leaves–which are whitish on both the upper and lower surfaces–and bark more deeply furrowed. At the upper limits of its range, notably on exposed ridges, western hemlock more closely resembles its hardy cousin, and hybrids have been reported but not confirmed.

Mountain Hemlock cones *Tsuga mertensiana*

Mountain Hemlock trunk *Tsuga mertensiana*

Mountain Hemlock, krummholz form *Tsuga mertensiana*

Douglas Fir *Pseudotsuga menziesii*

Douglas fir is one of the largest trees in the Pacific Northwest, often more than 250 feet (75 m) tall and 6 feet (2 m) in diameter. The crown is narrowly cone shaped, irregular, and has drooping branches. The bark is dark brown to black, very thick, and deeply furrowed.

Leaves: Dark to pale green, spreading around the twigs, pointed but not sharp, about 1 inch (2 to 3 cm) long.

Cones: Cylindrical, 2 to 4 inches (5 to 10 cm) long; conspicuous three-pointed bracts extend beyond the cone scales.

Ecology: Douglas fir grows on both sides of the mountain ranges from central British Columbia southward, at low to medium-high elevations. Although the trees grow larger and faster on the moist, western slopes, they are not as shade tolerant as hemlocks and true firs, which tend to replace them in old-growth forests. On the drier, eastern slopes, it is a climax dominant at mid-elevations.

Douglas fir is easy to identify. The cones have three-pronged bracts that extend beyond the cone scales, fancifully resembling the two hind legs and tail of a mouse diving into its hole. If cones are not on the tree, Douglas fir is distinguished from other conifers by its sharp-pointed, shiny, chestnut-colored buds.

Historically, fire has played a major role in the success of Douglas fir, especially on the western slopes of the mountain ranges. Fires kill the more competitive hemlocks and true firs, enabling the fast-growing Douglas firs to become established. Also, Douglas fir tolerates fire better than the hemlocks and firs. Because of its rapid growth and its adaptability and high-quality wood, Douglas fir is the most valuable lumber tree in the region.

Douglas Fir cone ***Pseudotsuga menziesii***

Douglas Fir trunk ***Pseudotsuga menziesii***

Fire in Douglas Fir forest

YEW FAMILY Taxaceae

The yews comprise a small family of evergreen trees and shrubs differing from other conifers because they bear seeds singly, instead of in cones. Only one species of this family grows in western North America.

Western Yew *Taxus brevifolia*

Western yew is a small to medium-sized, unisexual tree. The stems are crooked and extensively branched; the bark is thin and reddish. In recent years, western yew has been harvested–and overharvested–as a source for taxol, a drug used to treat cancer. Otherwise, because the trees are not particularly large and the stems are crooked, it has little commercial value. Yew trees produce wood that is notably tough and resilient though, and is renowned for making fine longbows.

Leaves: Needlelike, typically arched downward, somewhat paler on the lower surface, with a short stalk (petiole) and a fine, soft point.

Cones: Instead of cones, yews produce single-seeded, berrylike fruits with a red pulpy covering; reputed to be poisonous.

Ecology: Western yew grows primarily in old-growth forests, mainly along the lower montane zone more or less throughout the region; seldom is it common.

Western Yew ***Taxus brevifolia***

Western Yew trunk ***Taxus brevifolia***

Western Yew berry ***Taxus brevifolia***

Hardwoods (Angiosperms)

The hardwoods or broad-leaf deciduous trees of the Pacific Northwest primarily grow as seral species in conifer-dominated forests or in specialized habitats, such as along waterways. Angiosperms are flowering plants, although among most Northwest hardwoods the flowers are inconspicuous. Hardwoods, like all angiosperms, produce seeds in ovaries rather than cones, and some hardwoods produce single-seeded fruits called achenes.

Pine and maple autumn foliage

Birch Family — Betulaceae

The birch family is small, composed of trees and large shrubs with alternate leaves and branches. The flowers are borne in catkins, both male and female on the same plant. The leaves are conspicuously toothed or lobed and are nearly as wide as long. See page 413 for illustrations of birch family catkins and flowers.

Red Alder — *Alnus rubra*

Red alder is a medium-sized tree, up to 100 feet (30 m) tall and more than 2 feet (60 cm) in diameter. The bark is smooth, pale green to reddish, and typically mottled by white patches of lichen.

Leaves: Alternate, broadly elliptical, 2 to 5 inches (6 to 12 cm) long, dark green above and paler beneath, with conspicuous sunken veins; leaf margins have coarse, rounded teeth and are slightly rolled under.

Flowers: Inconspicuous and densely clustered in catkins that appear in spring before the leaves; female catkins woody, conelike, and persistent on the branches; male catkins long, caterpillar-like, and pendent.

Fruits: Single-seeded, winged achenes; several produced in each woody catkin.

Ecology: Well known as a fast-growing tree that invades logged, burned, and other open areas, including unstable river bars; ultimately it is replaced by more shade-tolerant conifers; ranges from Alaska south through Washington, mainly on lower, west-facing mountain slopes. During its short stay (fifty to seventy-five years), it enriches the soil through nitrogen fixation.

Sitka Alder — *Alnus sinuata*

This alder grows as a small tree or large shrub, usually with several spreading stems. The bark is smooth and dark green.

Leaves: Alternate, egg shaped, pointed, sharply and irregularly toothed, 2 to 4 inches (4 to 10 cm) long; veins conspicuous and sunken.

Flowers: Inconspicuous and densely clustered in catkin resembling those of red alder but appearing with, rather than before, the leaves.

Fruits: Single-seeded, winged achenes; each woody catkins produces several achenes.

Ecology: Prefers moist, open areas and often forms dense, nearly impenetrable thickets in areas of deep snowpack in the subalpine zone along avalanche tracks; common throughout the Pacific Northwest, from lowlands upward to the treeline.

Red Alder *Alnus rubra*

Red Alder trunk *Alnus rubra*

Sitka Alder *Alnus sinuata*

Sitka Alder *Alnus sinuata*

Paper Birch

Betula papyrifera

Paper birch is usually a medium-sized tree but can grow to more than 125 feet (38 m) tall and 3 feet (1 m) in diameter. The bark is typically white, sometimes reddish, thin, and smooth but conspicuously marked by black horizontal lines.

Leaves: Alternate, egg shaped, 1 to 3 inches (3 to 8 cm) long, doubly sharp-toothed (teeth on teeth).

Flowers: Densely clustered and inconspicuous in catkins; female catkins have deciduous scales, not woody and persistent as in alders; male catkins pendant and caterpillar-like.

Fruits: Small, single-seeded, winged achenes.

Ecology: A common associate of white spruce in the boreal forests of Canada and Alaska, extending southward into northern Washington, where it does well in moist sites on both sides of the Cascades.

Western birch *(B. occidentalis)* grows on the eastern side of the Cascades in Washington and the Coast Mountains of British Columbia, mainly along streams. Leaves are singly, not doubly, toothed and the bark is copper colored. Apparently it hybridizes with paper birch where the ranges of the two species overlap.

Bog Birch

Betula glandulosa

Sometimes called swamp birch, bog birch grows as a small tree or shrub, usually less than 10 feet (3 m) tall, with reddish brown bark. The branches are densely hairy and covered with conspicuous, crystal-like (glandular) warts.

Leaves: Alternate, broadly elliptical, ¼ to 1½ inches (1 to 4 cm) long, finely toothed; petioles warty.

Flowers: Densely clustered in catkins similar to those of paper birch.

Fruits: Small, single-seeded, winged achenes.

Ecology: Ranges from northern Washington, where it is uncommon, into Alaska; prefers wet habitats, such as bogs, but in northern British Columbia and adjacent Alaska, it often forms dense communities in moist, subalpine meadows.

Paper Birch *Betula papyrifera*

Paper Birch trunk *Betula papyrifera*

Bog Birch *Betula glandulosa*

Maple Family — Aceraceae

The maple family exhibits a combination of characteristics that include opposite leaves and branches, palmately lobed or divided leaves, and winged fruit (called a samara) that eventually divides into two one-seeded units. The flowers vary among species, from highly reduced and often unisexual to rather showy. The sepals and petals are usually similar in color and size.

Vine Maple — *Acer circinatum*

The most shade-tolerant maple in the Pacific Northwest, this is a small tree or large shrub, typically with several spreading and sprawling stems (trunks) that are extensively branched. The branches root freely when they come in contact with the soil, resulting in a dense tangle. The bark is green and smooth.

Leaves: Opposite, with seven to nine shallow lobes.

Flowers: Borne in small clusters along the branches; sepals red.

Fruits: Usually reddish, the wings spreading at nearly 180 degrees.

Ecology: Often grows in the conifer forest understory, particularly in old-growth forests; also spreads onto open slopes, where it becomes conspicuous in autumn with colorful red to orange leaves; ranges from lowlands to the subalpine zone in Washington and the Coast Mountains of southern British Columbia.

Douglas Maple — *Acer glabrum* var. *douglasii*

Like vine maple, Douglas maple may be either a small tree or large shrub, usually with multiple, more or less erect stems with reddish, smooth bark.

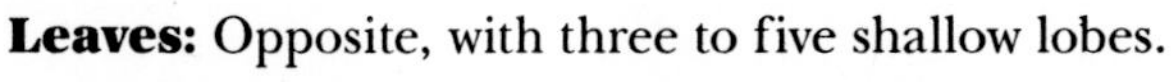

Leaves: Opposite, with three to five shallow lobes.

Flowers: Borne in clusters of five to ten; sepals and petals green.

Fruits: Green to reddish, the wings spreading at about 135 degrees.

Ecology: Grows along the Pacific coast and on the eastern slopes of the Cascades in Washington and southern British Columbia. When sufficiently exposed to sunlight on open ridges, the leaves turn brilliant red and orange in autumn.

Vine Maple fruits *Acer circinatum*

Vine Maple autumn foliage *Acer circinatum*

Douglas Maple *Acer glabrum*

Big-Leaf Maple *Acer macrophyllum*

Big-leaf maple is usually a medium-sized tree, but can grow to be about 100 feet (30 m) tall and more than 3 feet (1 m) in diameter—many old trees with branched trunks have a much greater girth. On the drier, eastern side of Pacific Northwest mountain ranges, big-leaf maples are smaller and sometimes shrubby. The bark is gray to brown and shallowly fissured.

Big-leaf maple is long lived, and senescent trees persist in old-growth forests. Typically these trees are laden with mosses, lichens, and ferns hanging from the branches in sheets, which creates a ghostly atmosphere. Trees growing in the open, notably on the eastern side of the mountain ranges, are beautiful in autumn, when the leaves turn golden yellow.

Leaves: Opposite, with three major and two minor (basal) lobes, often very large, up to 10 inches (25 cm) wide.

Flowers: Borne in dense, elongate, hanging clusters that include both unisexual and bisexual flowers; sepals and petals yellow-green.

Fruits: Green to brownish, densely hairy, the wings spreading at about 90 degrees, more or less perpendicular to each other.

Ecology: A seral species at low to mid-elevations in the mountains, primarily on moist, west-facing slopes. Ranges from Alaska southward through Washington, but rare in Alaska and northern British Columbia.

Big-Leaf Maple *Acer macrophyllum*

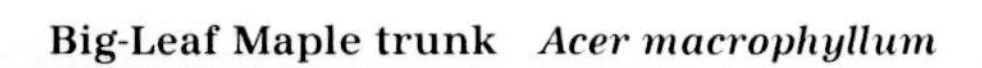

Big-Leaf Maple trunk *Acer macrophyllum*

WILLOW FAMILY

Salicaceae

This is a large family of trees and shrubs, with alternate, shallowly toothed or nontoothed (entire) leaves. The flowers are inconspicuous and borne in catkins, male and female on separate plants, meaning the plants are unisexual. The fruit matures into a capsule with numerous tiny seeds tufted with "cotton" that facilitates dispersal by the wind.

Quaking Aspen

Populus tremuloides

Aspen is a medium-sized tree, up to 75 feet (23 m) tall and 2½ feet (75 cm) in diameter, with distinctive bark that is thin, smooth, and pale green to white.

Quaking aspen is a beautiful tree in all seasons, particularly in autumn, when the foliage turns golden yellow. Its primary method of reproduction is by underground rhizomes. A recent study concluded that the largest single organism, in total biomass, was a clone of aspen covering several acres.

Leaves: Alternate, broadly egg shaped, the blade 1 to 2 inches (3 to 5 cm) long and just as wide; leaf stalk is flattened, which explains the leaves' "quaking" motion in a breeze.

Flowers: Densely clustered and individually indistinct in hairy, pendant, unisexual catkins.

Fruits: Capsules that split open at maturity, releasing numerous tufted seeds to blow in the wind.

Ecology: Although quaking aspen grows on the western side of the mountains, usually at low elevations, it is much more common and typically much larger on drier eastern slopes, where it extends upward from the sagebrush steppe to near the treeline; prefers stream banks and rich soil deposits at the base of mountain drainages; often forms expansive stands across mountain slopes, where it contrasts sharply with associated conifers; ranges from Alaska southward through Washington.

Aspen may be confused with paper birch because they both have whitish bark. They can be differentiated by examining the bark closely. Thin horizontal lines scar the bark of birch, while aspen has prominent black "eyes," which are scars where branches once grew.

Quaking Aspen autumn coloration ***Populus tremuloides***

Quaking Aspen (female) catkins ***Populus tremuloides***

Quaking Aspen trunk ***Populus tremuloides*** –George W. Douglas photo

Black Cottonwood *Populus balsamifera* ssp. *trichocarpa*

Often a very large tree, black cottonwood may grow to 175 feet (53 m) tall and 5 feet (1.5 m) in diameter. The bark is dark gray to black, rough, and deeply fissured. The winter buds are large, resinous, and pleasantly fragrant, particularly in spring, when the leaves are appearing. In autumn, black cottonwood provides a beautiful golden display along river banks. In spring, millions of seeds are released from each tree, creating what looks like a snowstorm as they float down from the sky and collect on the forest floor.

Leaves: Alternate, egg shaped to broadly triangular, tapering to a sharp point; the blade often exceeds 2 inches (5 cm) in length and width; the lower surface may be pale or covered by rust-colored hair.

Flowers: Densely clustered in unisexual catkins resembling those of aspen but larger, sticky, and nonhairy at maturity; male catkins red from the numerous anthers.

Fruits: Capsules, several per catkin, that release thousands of minute, cottony-tufted seeds to blow in the wind, thus the name *cottonwood.*

Ecology: Ranges from Alaska southward through Washington, on both sides of the mountain ranges; most frequently found along the major rivers, where it achieves maximum size.

Balsam poplar *(P. balsamifera* ssp. *balsamifera)* closely resembles black cottonwood but is smaller, and the female flowers have two stigmas rather than three. The two subspecies overlap in distribution and hybridize along the river drainages on the eastern side of the Coast Mountains in northern British Columbia.

Willows *Salix* species

Although some willows are large and treelike, for convenience and to avoid repetition, willows (*Salix* species) are treated together in the Forbs and Shrubs section on pages 348 to 357.

Black Cottonwood male catkins
Populus balsamifera

Black Cottonwood trunk
Populus balsamifera

Black Cottonwood riparian community *Populus balsamifera*

Buckthorn Family Rhamnaceae

Only one tree species in this small and variable family grows in the Pacific Northwest.

Cascara *Rhamnus purshiana*

Cascara is a small tree or large shrub, up to 30 feet (9 m) tall, with thin, smooth, pale gray bark. Cascara is best known for its bark, which historically has been collected and used as a laxative. However, the plants are not particularly common, even though the seeds are widely dispersed by birds and seedlings are abundant along the forest floor. The plants cannot tolerate deep shade and seldom reach maturity.

Leaves: Alternate, though nearly opposite, elliptical, 2 to 5 inches (5 to 12 cm) long, glossy green with conspicuous, sunken, pinnate veins.

Flowers: Greenish white, borne in small, branched clusters near the branch tips.

Fruits: Purplish black berries, bitter but edible.

Ecology: An occasional tree in mixed forests at low to mid-elevations, mainly on the western slopes of the Cascades in Washington and adjacent British Columbia.

A distantly related shrub in the buckthorn family is **mountain balm** *(Ceanothus velutinus),* named for the pleasant odor of the foliage. This shrub has sticky, evergreen, conspicuously veined leaves and dense clusters of small, white flowers. It grows in open Douglas fir and ponderosa pine forests on the eastern slopes of the Cascades in Washington and adjacent British Columbia.

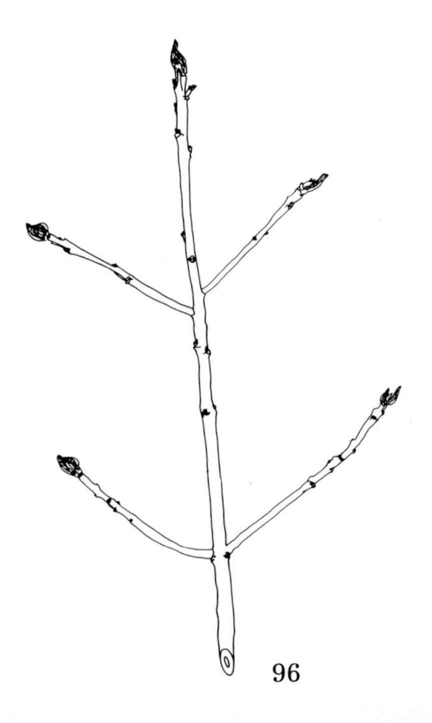

Cascara *Rhamnus purshiana* –George W. Douglas photo

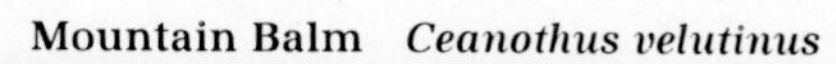

Mountain Balm *Ceanothus velutinus*

Rose Family

Rosaceae

Although this family is described and treated in the Forbs and Shrubs section (pages 262 to 284), it includes one tree best described here. **Bitter cherry** *(Prunus emarginatus)* grows to be about 30 feet (9 m) tall. The bark is thin, reddish brown to gray, and has horizontal markings similar to the lines on paper birch. The leaves are deciduous, alternate, elliptical, 1 to 3 inches (3 to 8 cm) long, and finely toothed. The flowers are white, about ½ inch (12 mm) across, and grouped in flat-topped clusters. The fruit is a bitter, bright red cherry. Bitter cherry is common in the mountains of Washington and southern British Columbia, in logged and other open areas from lowlands to moderately high elevations.

Bitter Cherry *Prunus emarginatus*

FORBS AND SHRUBS

The Forbs and Shrubs section has descriptions of the majority of the plants treated in this book and includes the most common wildflowers. The plants are arranged alphabetically by family; families can be identified by the Forbs and Shrubs key located in the back of the book. Drawings showing the anatomy of flowers and leaves are located in the Glossary.

Arum Family

Araceae

The arums comprise a large but primarily tropical family, characterized by small flowers sunken in a fleshy spike (spadix) and an associated, usually showy, modified leaf (spathe). The family is noted for its strong skunky or carrion-like odor, which is given off when the flowers are mature. The unpleasant odor attracts flies, which pollinate the flowers.

Skunk Cabbage

Lysichitum americanum

Skunk cabbage is a perennial herb with a thick underground stem.

Leaves: Large and cabbagelike, up to 3 feet (1 m) long and half as wide, glabrous and shiny.

Flowers: Small and numerous, greenish, sunken in the fleshy spike partially ensheathed by a bright yellow spathe.

Fruits: Leathery, remaining closed when mature (indehiscent), one or two seeds.

Ecology: A common plant of swampy or boggy habitats throughout the Pacific Northwest, from lowlands to mid-elevations (in the Cascades); especially showy in early spring before the leaves develop, when the lanternlike spathes are "flashing."

Birthwort Family

Aristolochiaceae

Birthworts make up a small family with the characteristics of wild ginger.

Wild Ginger

Asarum caudatum

Wild ginger is a trailing herb with stems that root at the nodes. Its flowers are inconspicuous on the forest floor and attractive, with an unusual coloration. The rhizomes taste like ginger.

Leaves: Opposite, blades broadly heart shaped, 3 to 5 inches (8 to 12 cm) wide, dark green and conspicuously veined.

Flowers: Sepals three, petal-like (petals absent), brownish purple, with a threadlike tip; stamens twelve.

Fruits: Many-seeded pods (capsules).

Ecology: A common but inconspicuous plant of moist forests, especially along streams, ranging from southern British Columbia through Washington.

Skunk Cabbage *Lysichitum americanum*

Skunk Cabbage *Lysichitum americanum*

Wild Ginger *Asarum caudatum*

BARBERRY FAMILY Berberidaceae

The nontechnical characteristics that loosely link the members of this diverse family are the flowers, which have multiple whorls of sepals or petals, the berrylike fruit, and the pinnately or ternately compound leaves.

Vanilla Leaf *Achlys triphylla*

Vanilla leaf is an herb lacking a leafy stem. It spreads by rhizomes.

Leaves: Trifoliate, with long stalks or petioles borne from the underground rhizome; blades up to 8 inches (20 cm) broad, three leaflets fan shaped and coarsely toothed.

Flowers: Individually very small, white, and densely clustered in a spike, sepals and petals alike, narrowly wedge shaped.

Fruits: Reddish, one-seeded berries.

Ecology: A common plant in moist, coniferous forests of western Washington and southern British Columbia, mainly at lower elevations in the mountains.

Oregon Grape *Berberis* species

Oregon grape is a low to medium-sized shrub.

Leaves: Evergreen, shiny, and leathery; alternate and pinnately compound, leaflets spiny along the margin, hollylike.

Flowers: Yellow, in narrow racemes; sepals and petals similar, usually in five whorls, three per whorl; stamens fused to the base of the inner petals.

Fruits: Blue to purple berries.

Three species of Oregon grape grow in the Pacific Northwest. On the western side of the Cascades, extending into British Columbia, the **Cascade Oregon grape** *(B. nervosa)* is common, especially in dry forests at low to mid-elevations. This species has nine or more palmately veined leaflets per leaf. On the eastern side of the Cascades, **creeping Oregon grape** *(B. repens)* is common, extending upward from the sagebrush steppe to high rocky ridges. It has five to seven, pinnately veined leaflets. **Tall Oregon grape** *(B. aquifolium)* is less common but grows on both sides of the Cascade Range in Washington and southern British Columbia. It is taller than the other species, reaching more than 2 feet (60 cm) high. The berries of all species are edible and provide a valuable food source for birds and other animals.

Vanilla Leaf *Achlys triphylla*

Tall Oregon Grape *Berberis aquifolium*

Cascade Oregon grape *Berberis nervosa*

Bladderwort Family Lentibulariaceae

This is a small family of insectivorous plants. In one of two genera inhabiting the Northwest, *Utricularia,* modified leaves form "traps" that catch insects; in the other, *Pinguicula,* leaf surfaces are sticky, and insects simply become glued to them. The leaves then secrete enzymes that digest the insects. Bladderwort flowers are similar to those of the closely related figwort family.

Butterwort *Pinguicula vulgaris*

A mountain plant, butterwort supplements its nitrogen requirement by feeding on insects. It is a small, perennial herb with attractive flowers and a single, unbranched leafless stem, up to 6 inches (15 cm) high.

Leaves: Basal, succulent, and slimy (sticky), 1 to 2 inches (2 to 5 cm) long.

Flowers: Bilaterally symmetrical, blue, petals fused into a tube with a long spur and two lips.

Fruits: Many-seeded capsules.

Ecology: A bog plant that grows sporadically throughout the region; largely replaced in the north by *P. villosa.*

Buckbean Family Menyanthaceae

The diagnostic features of this family are mainly technical, relating to the flower structure. In general, the petals are fused into a tube with fringed or bearded lobes, the several flowers are borne on leafless stalks, and the leaves are all basal.

Deer-Cabbage *Fauria crista-galli*

Attractive and uncommon, deer-cabbage is a succulent, perennial herb.

Leaves: All basal, blades round to broadly heart shaped, coarsely toothed, glabrous, and shiny.

Flowers: Petals white, fused into a tube with five spreading and fringed lobes, stamens opposite the petal lobes.

Fruits: The pistil becomes a many-seeded pod (capsule).

Ecology: Ranges from Alaska to northwestern Washington, mainly along the coast in bogs and wet meadows.

Buckbean *(Menyanthes trifoliata)* is a related bog species in the same region.

Butterwort *Pinguicula vulgaris*

Dear-Cabbage *Fauria crista-galli*

Buckbean *Menyanthes trifoliata*

Borage Family Boraginaceae

The borage family has several characteristics that distinguish it from other plant families. The five petals are fused at the base to form a tube that conceals the nectar, preserving it for long-tongued insects such as butterflies and some bees. At the mouth of the petal tube, five flaplike structures are associated with the petal lobes. The five stamens are fused to the inside of the petal tube. The pistil is four-lobed, each lobe maturing into a one-seeded nutlet. The flower cluster (inflorescence) is usually coiled, and the plants are often densely hairy. The family is best represented in deserts and steppes, but many species grow in the high mountains.

Alpine Forget-Me-Not *Eritrichium nanum*

Undoubtedly one of the most attractive alpine wildflowers, alpine forget-me-not is a stereotypical alpine cushion plant. It has numerous short, erect flowering stems, 2 to 6 inches (5 to 15 cm) high and is a perennial herb.

Leaves: Alternate and basal, cylindrical to lance shaped, less than ½ inch (12 mm) long, covered with long hair.

Flowers: Pale to dark blue with a white to yellow ring around the throat of the tube, 4 to 8 mm across; other floral characteristics are typical of the family.

Fruits: Four angular, hairless nutlets per flower.

Ecology: Primarily a Rocky Mountain species, but occurs infrequently on rocky, alpine ridges in the North Cascades of Washington, on the eastern side of the crest.

Mountain forget-me-not *(Myosotis alpestris)* is a similar plant, but it is much taller and does not form mats. The stems are low, not more than 1 foot (30 cm) high, and clustered, with several per plant. The leaves are longer than wide, the lower ones stalked and the upper ones reduced in size and lacking petioles. The blossoms are sky blue to dark blue, in coiled clusters at the top of hairy stems.

This beautiful wildflower is abundant in Alaska, where it is the state flower. It extends southward to southern British Columbia in moist, subalpine meadows and along streams in open forests.

Alpine Forget-Me-Not *Eritrichium nanum*

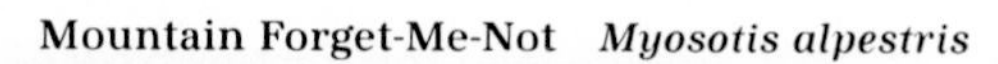

Mountain Forget-Me-Not *Myosotis alpestris*

Blue Stickseed *Hackelia micrantha*

A perennial herb with multiple branched stems, blue stickseed grows up to 3 feet (1 m) tall. Stiff hairs clothe the stem and leaves.

Leaves: Alternate, more or less lance shaped, lower leaves with long petioles, upper leaves reduced and stalkless (sessile).

Flowers: Sky blue to lavender, 5⁄16 to 3⁄8 inch (8 to 10 mm) across; other floral characteristics are typical of the family.

Fruits: Four nutlets with many hooked bristles that aid in dispersal by animals. Stickseed is an appropriate common name for this plant.

Ecology: Locally common in dry, subalpine meadows; occasionally grows in open forests; ranges from southern British Columbia through the mountain ranges of Washington, primarily on the dry, eastern slopes.

Tall Bluebell *Mertensia paniculata*

Bluebells are succulent, herbaceous perennials. Tall bluebell has stems up to 4½ feet (1.5 m) high, usually several per plant.

Leaves: Basal leaves ovate to heart shaped, with long petioles; stem leaves abundant, alternate, lance shaped, up to 6 inches (15 cm) long, becoming progressively smaller higher on the stem.

Flowers: Blue to lavender, about ½ inch (12 mm) long, bell shaped and pendant; petals fused into a tube, the terminal part (the limb) enlarged with five short lobes, longer than the narrow base of the tube.

Fruits: Four small, rough nutlets.

Ecology: An occasional plant of moist forests, stream banks, and high-elevation meadows; ranges from Alaska through Washington, mainly on the eastern side of the Cascades.

Two shorter species of *Mertensia* also grow in the Pacific Northwest, *M. longiflora* and *M. oblongifolia,* both called **low bluebell.** They have narrower, less bell-shaped flowers, and the terminal limb of the petal tube is much shorter than the narrower basal component. Both species inhabit the eastern side of the Cascades in Washington and Oregon; *M. longiflora* ranges north into southern British Columbia, and both extend east into the Rocky Mountains. These bluebells are common, early flowering wildflowers of the sagebrush steppe, extending upward in open forests and onto upland slopes and mountain ridges.

Blue Stickseed *Hackelia micrantha*

Tall Bluebell *Mertensia paniculata*

Tall Bluebell *Mertensia paniculata*

Low Bluebell *Mertensia oblongifolia*

Buckwheat Family Polygonaceae

Because of pronounced variation among the buckwheat species, this is a difficult family to characterize. Although the flowers lack petals, the sepals are often showy, though small, and petal-like. With exceptions, the species fall into one of two groups: (1) the **knotweeds,** typified by mountain bistort. These are herbs, usually with five sepals. A papery sheath surrounds the stem above the leaves, which taste like rhubarb. The flowers are often densely clustered but not in umbels. (2) The **buckwheats** are dwarf shrubs, with six sepals in two whorls of three. They lack sheathing stipules and do not taste like rhubarb. The flowers are borne in dense umbels.

Umbrella Buckwheat *Eriogonum umbellatum*

Umbrella buckwheat is a dwarf, matted shrub with short, erect flowering stems reaching upward about 4 inches (10 cm).

Leaves: Generally elliptical, ⅜ to ¾ inch (1 to 2 cm) long and half as wide, borne in whorls at the base of erect stems and below the flower cluster; leaf undersurface covered with white woolly hair.

Flowers: Cream colored to yellow with six sepals, small and densely clustered in a ball-shaped umbel.

Fruits: Each flower produces a single achene.

Ecology: Several varieties of umbrella buckwheat extend upward from deserts onto alpine ridges; the most common alpine varieties are ***hausknechtii,*** in the Washington Cascades, and ***subalpinum,*** in the mountains of southern British Columbia.

Other species of buckwheat also grow in the Cascades of Washington and southern British Columbia. Perhaps the two most common are **alpine buckwheat** *(E. pyrolifolium)* and **oval-leaf buckwheat** *(E. ovalifolium).* Alpine buckwheat grows from a taproot and has numerous spreading basal leaves that are woolly beneath. The flowers vary in color from white to reddish and are clustered in round umbels. Two elongate bracts at the base of each umbel mark the species. Oval-leaf buckwheat is another highly variable species, ranging from deserts upward into the alpine zone, where it forms dense cushions. Both the leaves and the short, leafless flowering stems are covered by white, woolly hair, an adaptation that protects the plants from blazing sun and drying winds. The dense ball of flowers varies in color from white to red (especially with age) or even yellow.

Umbrella Buckwheat *Eriogonum umbellatum*

Alpine Buckwheat *Eriogonum pyrolifolium*

Oval-Leaf Buckwheat *Eriogonum ovalifolium*

Mountain Sorrel *Oxyria digyna*

Mountain sorrel is a somewhat fleshy, perennial herb with a few to several erect stems, 6 to 24 inches (1 to 6 dm) tall. The entire plant is hairless. This species is in the same family as rhubarb and has the characteristic rhubarb flavor, thus the common name "sorrel."

Leaves: Mainly basal, with long stalks and broadly heart-shaped blades.

Flowers: Small, with four green to reddish sepals (no petals), densely clustered along the upper half of the erect stems.

Fruits: Each flower produces a single, one-seeded, winged achene.

Ecology: Widely distributed, ranging from the Arctic tundra southward along all major mountain ranges of North America, most frequently found in rock crevices and on scree slopes where moisture is abundant.

Mountain Bistort *Polygonum bistortoides*

Mountain bistort is a perennial herb with one to several erect flowering stems, 1 to 2½ feet (3 to 7.5 dm) tall.

Leaves: Mainly basal, with long petioles and elliptical blades; progressively higher on the plant, the stem leaves become smaller, more lance shaped, and lack petioles.

Flowers: Small, white to pink, five sepals, borne in a dense, round to elliptical cluster at the top of the stem; stamens extend beyond the petals, giving the flower cluster a brushlike appearance.

Fruits: A triangular, single-seeded achene.

Ecology: One of the most common and conspicuous residents in subalpine meadows of western North America, but in the Pacific Northwest, it grows only in Washington and southern British Columbia.

This is a variable species. The more northern plants tend to be smaller and are more likely to have pink rather than white flowers. A less attractive, locally common species in the southern Cascades and Washington Olympics is **Newberry's knotweed** *(P. newberryi).* This ungainly plant has branched, erect to prostrate, leafy stems. The small, greenish white to reddish flowers are partially hidden among the ovate to elliptical leaves. It grows on moist, rocky, subalpine and alpine slopes.

Mountain Sorrel *Oxyria digyna*

Mountain Bistort *Polygonum bistortoides*

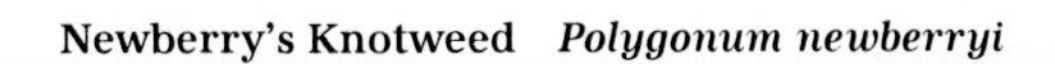

Newberry's Knotweed *Polygonum newberryi*

BUTTERCUP FAMILY Ranunculaceae

The buttercup family may be the most primitive family of flowering plants in the Northwest, closely related to fossil forms from the Cretaceous Period, more than 100 million years ago. The most obvious primitive characteristics relate to the flowers, which have numerous stamens, multiple pistils, and no fusion of floral parts. However, the family combines plants with primitive, nonspecialized flowers with plants that are adapted for pollination by bumblebees (larkspurs and monkshoods), hummingbirds (columbines), and wind (meadowrue). These specialized forms retain the primitive characteristics of numerous stamens, multiple pistils, and no fusion of parts. All plants in the buttercup family are nonwoody and occasionally viney; most of them have compound leaves.

Monkshood *Aconitum columbianum*

Monkshood is a coarse, branched, perennial herb, up to 6 feet (2 m) tall, with one to several hollow stems. Monkshood flowers are highly specialized for pollination by bumblebees. The bees manipulate the flowers to reach the nectar concealed in the upper sepal, which fancifully resembles the cowl or hood of a monk's garment, thus the common name. As is true of many members of the buttercup family, monkshood species contain poisonous alkaloids.

Leaves: Alternate, palmately compound with three to five major, variously toothed and divided segments; the larger, lower leaf blades may be more than 6 inches (15 cm) wide, upper leaves progressively smaller, becoming bractlike.

Flowers: Dark blue or occasionally white, borne in terminal racemes; bilaterally symmetrical with five sepals, the upper one modified into a pointed hood, the two lateral ones roundish, and the two lower ones lance shaped; two inconspicuous petals, largely hidden by the colorful sepals; numerous stamens; three to five pistils.

Fruits: The three to five pistils develop into many-seeded pods that split open at maturity.

Ecology: A common species in moist woods and along streams, particularly on the eastern slopes of the Cascades; ranges from northern British Columbia southward along mountain ranges, perhaps as far as Mexico.

In Canada and Alaska, this species is largely replaced by **mountain monkshood** *(A. delphinifolium),* a smaller plant of mountain meadows and tundra habitats, occasionally growing in moist, open forests.

Monkshood *Aconitum columbianum*

Monkshood *Aconitum columbianum*

Mountain Monkshood
Aconitum delphinifolium

Mountain Monkshood
Aconitum delphinifolium

Baneberry

Actaea rubra

This is a perennial herb with one or more stems, growing to 3 feet (1 m) tall. European species of *Actaea* are known to be highly poisonous, and the North American species is suspect. *Baneberry* means "murderous" or "destroying" berry.

Leaves: Usually two, the lower one much larger than the upper, generally divided three times, with variously toothed and lobed leaflets.

Flowers: Individual flowers small, sepals and petals similar in color (white) and size, much shorter than the numerous stamens; flowers borne in a dense, roundish cluster at the stem tip.

Fruits: Shiny red or white berries.

Ecology: An occasional plant in moist forests and along streams throughout the Pacific Northwest.

Oregon Anemone

Anemone oregana

Oregon anemone is a perennial herb that spreads from shallow, creeping rhizomes. The flowering stems are solitary and leafless up to the base of the single flower. The plants grow to about 1 foot (30 cm) high.

Leaves: Two or three occur in a whorl just below the flower; occasionally an additional leaf with a very long stalk grows directly from the rhizome; both types of leaves are trifoliate to palmately compound, the leaflets variously toothed and about 2 inches (5 cm) long.

Flowers: Solitary at the stem tip; sepals five or sometimes more, petal-like (petals absent), about ¾ inch (14 to 20 mm) long, usually blue to purplish, occasionally pink or white; stamens and pistils numerous.

Fruits: Somewhat elongate, one-seeded, hairy achenes.

Ecology: Locally very common on the eastern side of the Washington Cascades; prefers moist conifer forests but sometimes grows on open slopes.

Baneberry flowers *Actaea rubra*

Baneberry fruits *Actaea rubra*

Oregon Anemone *Anemone oregana*

Drummond's Anemone *Anemone drummondii*

A common and attractive cushion plant, this hairy, herbaceous species has several short stems, each with a single flower. A whorl of small (involucral) leaves is borne below the flower.

Leaves: Basal leaves have long stalks, with blades more or less cloverlike (triternate); stem (involucral) leaves are similar but lack stalks.

Flowers: White to purplish; five to nine sepals, ⅜ to ¾ inch (1 to 2 cm) long, petal-like (petals absent); stamens and pistils numerous.

Fruits: Several one-seeded, woolly achenes, collectively resembling a ball of wool at the tip of the erect stems.

Ecology: Most frequently found on scree and talus slopes in the alpine zone; flowers in mid-spring and ranges from Alaska through the Washington Cascades and Olympics.

Cut-Leaf Anemone *Anemone multifida*

Cut-leaf anemone is a hairy perennial herb, with clustered stems that may be as much as 2 feet (0.6 cm) high.

Leaves: The basal leaves have long petioles; leaf blades divided into several narrowly elliptical segments or leaflets; a whorl of leaves is borne midway up the stem—these leaves are similarly divided but not long stalked.

Flowers: White to purplish or occasionally reddish, sepals usually five, petal-like (petals absent); about ⅜ inch (1 cm) long; stamens and pistils numerous.

Fruits: Several silky to woolly achenes in a silky ball at the tip of erect stems.

Ecology: Widely distributed more or less throughout the Pacific Northwest, from dry, open forests to rocky alpine ridges.

Cut-leaf and Drummond's anemone are highly variable and difficult to distinguish. However, Drummond's anemone is generally shorter, more cushionlike, and less hairy, with leaves somewhat less divided. Also, cut-leaf anemone often has more than one flower per stem. Drummond's anemone is more strictly a subalpine or alpine species, but the two species overlap in elevation and habitat preference as well as geographical range.

Drummond's Anemone *Anemone drummondii*

Cut-Leaf Anemone flowers
Anemone multifida

Cut-Leaf Anemone fruits
Anemone multifida

Mountain Pasqueflower *Anemone occidentalis*

Mountain pasqueflower, also called western anemone, is an attractive plant that appears soon after snowmelt. Later in the season, the large, shaggy heads of achenes make the plants conspicuous in lush meadow vegetation. Also known as western anemone, this perennial herb is covered with long, soft hair. The stems are clustered from a stout root crown, and reach 2 feet (0.6 m) high. Each stem bears a single flower and a whorl of reduced (involucral) leaves below the flower.

Leaves: Basal leaves with long stalks and divided, parsleylike blades; stem (involucral) leaves reduced in size and nonstalked.

Flowers: White to purplish; sepals five, petal-like (petals absent), about 1 inch (2 to 3 cm) long; stamens numerous; pistils also numerous, styles becoming long and feathery.

Fruits: Numerous single-seeded achenes, each developing a long, feathery style at maturity, which aids in wind dispersal of the achene.

Ecology: Inhabits subalpine meadows from the southern half of British Columbia through the Washington Cascades and Olympics.

Narcissus Anemone *Anemone narcissiflora*

Narcissus anemone, also known as alpine anemone, is a perennial herb, with a few to several stems derived from a stout root crown; the stems are hairless or nearly so and may be up to 2 feet (0.6 m) high.

Leaves: Basal leaves long stalked and cloverlike (trifoliate), each of the three leaflets deeply divided into narrow, pointed segments; stem leaves very much reduced and borne in a whorl below the single flower.

Flowers: White, or bluish on the lower surface; sepals petal-like, five to nine, about 1 inch (2 to 3 cm) long; stamens and pistils numerous.

Fruits: Several hairless achenes.

Ecology: Widely distributed in the Arctic tundra, ranging southward to central British Columbia; primarily a plant of subalpine meadows and open forests.

Mountain pasqueflower and narcissus anemone are similar in size and habitat, but pasqueflower is much hairier, especially the achenes, and the two species are distributed on opposite ends of the region.

Mountain Pasqueflower *Anemone occidentalis*

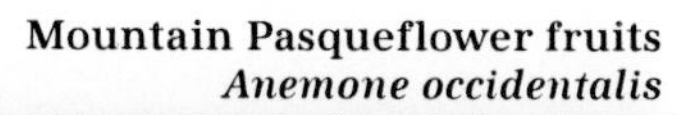

Mountain Pasqueflower fruits
Anemone occidentalis

Narcissus Anemone
Anemone narcissiflora

Northern Anemone *Anemone parviflora*

Northern anemone is a hairy, perennial herb that spreads by rhizomes. The stems usually grow singly, about 6 inches (15 cm) high.

Leaves: Basal leaves divided into three leaflets, which are wedge shaped and toothed at the tip, less that an inch (2.5 cm) long; the reduced stem leaves are borne in a whorl below the single flower.

Flowers: Solitary; sepals five, petal-like (petals absent), white to pale lavender, ⅜ to ⅝ inch (10 to 15 mm) long; stamens numerous; pistils several, densely hairy.

Fruits: Single-seeded achenes covered with long hairs.

Ecology: A common plant in the tundra of Alaska and northern British Columbia, extending southward in moist subalpine meadows and seepage areas to the Washington Cascades.

Another common anemone species in Alaska and northern British Columbia is **yellow anemone** *(A. richardsonii),* unusual among *Anemone* species because it has yellow flowers. Like northern anemone, it is a small, delicate plant but different because of its floral color and less deeply lobed leaves. It prefers wet habitats, mainly along streams and in seepages.

Marsh Marigold *Caltha leptosepala (biflora)*

Marsh marigold is an attractive, succulent, perennial herb. The plants are hairless and shiny and grow to about 6 inches (15 cm) high.

Leaves: Mainly basal, the one or two stem leaves reduced in size; leaf blade round to heart shaped or oblong, 2 to 4 inches (5 to 10 cm) long and about as wide, usually coarsely toothed; petiole longer than the blade.

Flowers: Usually two per plant; sepals six to twelve, petal-like (petals absent), whitish, about ¾ inch (2 cm) long; stamens numerous; pistils several.

Fruits: Each of the pistils matures into a many-seeded pod.

Ecology: Distributed throughout the Pacific Northwest, along streams and in seepage areas from montane forests upward into the alpine zone; frequently forms dense, strikingly attractive populations.

Northern Anemone *Anemone parviflora*
—George W. Douglas photo

Yellow Anemone *Anemone richardsonii*
—George W. Douglas photo

Marsh Marigold *Caltha leptosepala*

Red Columbine

Aquilegia formosa

The most widely distributed columbine, this is a beautiful perennial herb. (The Latin name *formosa* means "beautiful.") Several branched stems derive from a stout taproot. The spreading stems may be as much as 3 feet (1 m) long.

Leaves: Mainly basal, stem leaves alternate and progressively smaller from the base of the stem upward; basal leaves divided into multiples of three leaflets, which are further toothed or lobed.

Flowers: Very showy; sepals five, red, up to 1 inch (2.5 cm) long; petals five, differentiated into a yellow blade and reddish, nectar-producing spur, which is about as long as the sepals; stamens numerous, extending beyond the petals; pistils three to five. Typically, the flowers face downward (nod), with the spurs pointing upward.

Fruits: Each pistil matures into a many-seeded pod.

Ecology: Distributed throughout the Pacific Northwest, from moist, open lowland forests to subalpine meadows, where it sometimes grows in abundance.

On the eastern side of the Cascades in Washington and southern British Columbia, red columbine distribution overlaps **yellow columbine** *(A. flavescens),* which has somewhat smaller and uniformly yellow flowers. Where their ranges overlap, the two species freely hybridize because bumblebees and hummingbirds forage from flowers of both species. The result is complete intergradation and taxonomic chaos.

A third species, **blue columbine** *(A. brevistyla),* grows in British Columbia and southeastern Alaska. Like the other two species, this is a perennial with a taproot and many compound basal leaves. It is smaller in all respects, however. The sepals and spurs are pale blue with whitish petal blades. The spurs are hooked rather than straight as in the yellow and red forms. It grows in mountain meadows and in moist, open forests, often along streams. Although its flowers differ in size, color, and shape from those of the red and yellow columbines, it may hybridize with either of those species, again as a result of the indiscriminate foraging behavior of bumblebees.

Red Columbine *Aquilegia formosa*

Red Columbine *Aquilegia formosa*

Yellow Columbine *Aquilegia flavescens*

Pale Larkspur *Delphinium glaucum*

Pale larkspur is a perennial herb with a few to several coarse, hollow stems that reach heights of up to 5 feet (1.5 m).

Leaves: Alternate and basal, numerous, stem leaves smaller higher up, becoming bractlike in the inflorescence. Basal leaves have long stalks and large, palmately lobed and divided blades sometimes as much as 8 inches (20 cm) wide.

Flowers: Bilaterally symmetrical, numerous in dense racemes; sepals five, pale to dark blue-purple, up to ½ inch (12 mm) long, the nectar-containing spur about as long as the sepal blades; petals four, pale blue; stamens numerous; pistils three to five.

Fruits: Each pistil develops into a many-seeded pod.

Ecology: An occasional plant of the Cascades and Olympics in Washington, but more common in northern British Columbia and southeastern Alaska. It grows along gravelly streams in open forests and subalpine meadows.

Menzies' Larkspur *Delphinium menziesii*

Larkspurs are notoriously poisonous and frequently cause livestock deaths on ranges where suitable forage plants are sparse. Menzies' larkspur, the most common larkspur in the region, is a hairy perennial herb, usually with a single, sparingly branched stem that stands up to 2 feet (0.6 m) high. The stem is weakly attached to a tuberous root system.

Leaves: Alternate and basal, palmately divided and lobed, more or less round in outline; upper stem leaves smaller, becoming bractlike.

Flowers: Generally similar to the flowers of pale larkspur but somewhat larger, few to several in an open raceme, the lower ones with long pedicels.

Fruits: Three to five many-seeded pods.

Ecology: Grows on the western slopes of the Washington Cascades and Coast Mountains in British Columbia, preferring forest openings and subalpine meadows, especially on rocky substrates.

A similar species, **common larkspur** *(D. nuttallianum),* grows on the drier, eastern slopes of Pacific Northwest mountain ranges, extending downward into the sagebrush steppe. Common larkspur is distributed from southern British Columbia southward through Washington.

Pale Larkspur *Delphinium glaucum*

Menzies' Larkspur *Delphinium menziesii*

ommon Larkspur *Delphinium nuttallianum*

Common Larkspur *Delphinium nuttallianum*

Subalpine Buttercup *Ranunculus eschscholtzii*

Subalpine buttercup is a very attractive, more or less succulent, perennial herb with hairless, shiny stems and leaves. The flowering stems are erect but short.

Leaves: Variable, mainly basal, palmately or ternately divided, the segments wedge shaped and typically three lobed; stem leaves divided into elongate, narrow segments, often reduced and bractlike.

Flowers: Borne singly at stem tips; petals usually five (sometimes more), up to ¾ inch (2 cm) long and equally wide, bright, shiny yellow, much larger than the green sepals; stamens and pistils numerous.

Fruits: Numerous single-seeded achenes with a short, usually straight style.

Ecology: An extremely variable and widely distributed species, growing in some varietal form over much of western North America, including southeastern Alaska; prefers moist habitats and most frequently inhabits the gravelly banks along small streams or in rocky seepage areas in the subalpine and alpine zones.

Western Buttercup *Ranunculus occidentalis*

Western buttercup is a perennial herb with a few to several weak and spreading stems 6 to 18 inches (15 to 45 cm) long. The entire plant is usually sparsely covered with soft hair.

Leaves: Mainly basal with long petioles, triternate, each of the three leaflets further divided and toothed; stem leaves alternate, smaller, less divided, and nonstalked higher on the stem.

Flowers: Similar to but smaller than those of subalpine buttercup, few to several borne on the branching stems; sepals conspicuously hairy and bent backward (downward).

Fruits: Several single-seeded achenes with a short, slightly hooked style.

Ecology: Primarily found on coastal bluffs but occasionally extends upward into subalpine meadows, especially in British Columbia and southeastern Alaska.

Several additional buttercup species inhabit the Pacific Northwest, but most of them are weedy plants of low elevations. A common native species in moist, open forests is **little buttercup** *(R. uncinatus),* a plant with large, thrice-divided leaves and small, inconspicuous flowers; the petals are only about 3 mm long, about ⅛ inch.

Subalpine Buttercup *Ranunculus eschscholtzii*

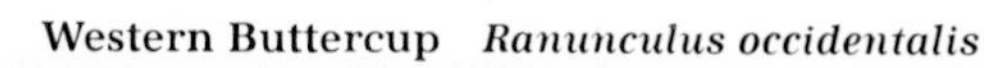

Western Buttercup *Ranunculus occidentalis*

Globeflower *Trollius laxus*

An early flowering plant, globeflower is a low, hairless, perennial herb with more or less succulent stems and leaves and shallow, fibrous roots.

Leaves: Basal and alternate, palmately divided, the segments or leaflets deeply toothed at the tip; basal leaves with long stalks, stem leaves mainly stalkless.

Flowers: Solitary at the tip of erect stems; sepals five or more, petal-like (petals absent), white to greenish, about ¾ inch (2 cm) long; stamens numerous; pistils several.

Fruits: Pistils mature into many-seeded pods.

Ecology: Inhabits wet, subalpine meadows and stream banks, from southern British Columbia through Washington.

Except for its divided leaves, globeflower resembles marsh marigold. The two species grow in similar habitats.

Western Meadowrue *Thalictrum occidentale*

A common, attractive herb, meadowrue is a unisexual perennial, up to 3 feet (1 m) tall, with branched stems. The entire plant is usually hairless.

Meadowrues exemplify adaptation to wind pollination. The plants are unisexual; the stamens have long, threadlike filaments enabling the large anthers to "rattle" in the wind and disperse the pollen; the pistils have feathery styles that comb the air for pollen; and the flowers are above the leaves and in the windstream.

Leaves: Alternate, mostly three times divided, with the leaflets also three lobed.

Flowers: Male and female flowers borne on separate plants; both types have five small green sepals and no petals; female flowers have several reddish pistils and male flowers have numerous pendant stamens, which make the flowers look like miniature chandeliers.

Fruits: Elongate, conspicuously veined achenes.

Ecology: A plant of subalpine meadows in British Columbia and Washington.

Other meadowrue species grow in the Pacific Northwest, mainly in drier habitats on the eastern slopes of the mountain ranges. Some of these have "perfect" (bisexual) flowers. All are similar in appearance.

Globeflower *Trollius laxus*

Western Meadowrue, male
Thalictrum occidentale

Western Meadowrue, female
Thalictrum occidentale

Currant Family Grossulariaceae

Usually, members of this family are grouped in a single genus, *Ribes,* and share many diagnostic characteristics. The plants are shrubs with alternate, palmately lobed or divided leaves, like maple leaves. Most species have a curranty odor, at times very strong. Many gooseberries have spiny stems. The sepals are larger and more colorful than the petals and are fused at the base to form a saucer-shaped or tubelike structure (calyx tube), from which other floral parts are borne. The ovary is situated below the other flower parts (inferior) and matures into a berry, edible but not always palatable.

Stink Currant *Ribes bracteosum*

Stink currant is an erect, nonspiny shrub, often more than 6 feet (2 m) tall, covered by glandular hairs that give the plant a pronounced, disagreeable odor.

Leaves: Alternate, deciduous, the blade large, up to 8 inches (20 cm) wide; palmately divided with five major, pointed lobes.

Flowers: Greenish purple, more or less saucer shaped; borne in narrow, erect or spreading racemes.

Fruits: Blue to black berries, covered with glandular hairs and unpalatable.

Ecology: A common shrub along streams and in wet forested areas, from lowlands to high elevations from Alaska through Washington.

In addition to maple-leaf currant, prickly currant, and sticky currant described here, several other species grow in some part of the Pacific Northwest. In Alaska and on the eastern side of the coastal mountains in northern British Columbia, **northern black currant** *(R. hudsonianum)* is particularly common. It has white flowers and black berries, resembles stink currant in form, size, and ecology, and has smelly glands. Its leaves are maplelike, with three to five major lobes.

Red swamp currant *(R. triste)* is also common in Alaska and northern British Columbia, much less so farther south into the Washington Cascades. A low, spreading, nonspiny shrub, it is seldom more than 3 feet (1 m) tall. Its leaves are shallowly three to five lobed, and the lobes are toothed. Red swamp currant flowers are saucer shaped, dark red, and borne in short, drooping racemes. The berries are bright red and palatable, but tart. As the common name suggests, this currant prefers moist to wet habitats, mainly along forest streams.

Red flowering currant *(R. sanguineum)* is an attractive shrub in the foothills of southern British Columbia and Washington. It has shallowly lobed and toothed leaves, red tubular flowers with spreading sepals (petals white), and bluish, glandular, unpalatable berries.

Stink Currant *Ribes bracteosum*

Red Flowering Currant *Ribes sanguineum*

Red Swamp Currant *Ribes triste*

Maple-Leaf Currant *Ribes howellii*

Common and sprawling, this high-mountain currant is an extensively branched, nonspiny shrub, usually less than 6 feet (2 m) tall.

Leaves: Alternate, deciduous, shallowly three to five lobed.

Flowers: Saucer shaped, borne in loose, hanging racemes; sepals white, usually mottled with red; petals dark red.

Fruits: Blue to black berries, covered with glandular hairs and unpalatable.

Ecology: Common in Washington, extending northward into southern British Columbia; frequently grows around tree clumps and along the treeline, usually in areas of deep snowpack.

Prickly Currant *Ribes lacustre*

A spreading shrub, prickly currant has flexuous branches and stems covered with prickles and stout spines.

Leaves: Alternate, deciduous, deeply palmately divided into three to five major, pointed segments, which are further toothed.

Flowers: Saucer shaped, dull yellowish green to reddish mottled, borne in hanging racemes.

Fruits: Black berries covered with glandular hairs but palatable.

Ecology: A common species in moist forests and along stream banks from Alaska to California, from lowlands to high in the mountains.

Sticky Currant *Ribes viscosissimum*

Sticky currant is a medium-sized, nonspiny shrub with straggly but stiff branches. Soft, sticky hairs cover the leaves and flowers.

Leaves: Alternate, deciduous, shallowly three to five lobed, toothed, glandular-sticky on both surfaces.

Flowers: Greenish white to yellow, often mottled with red, tubular at the base, covered with glandular-sticky hairs.

Fruits: Blue to black berries, glandular-sticky and unpalatable.

Ecology: A common shrub on the eastern side of the Washington Cascades and Coast Mountains in southern British Columbia. Grows in forests from middle to high elevations and around tree clumps in the subalpine zone.

Maple-Leaf Currant *Ribes howellii*

Prickly Currant *Ribes lacustre*

Sticky Currant *Ribes viscosissimum*

Dogwood Family

Cornaceae

Although dogwoods make up a large and variable family, only three species grow in the Pacific Northwest–bunchberry, flowering dogwood, and red osier dogwood. Of the three, only bunchberry is well represented in the mountains. The most obvious family characteristics relate to the leaves (either opposite or whorled and conspicuously veined), the flowers (parts in fours—four sepals, four petals, and so on); the inflorescences (either umbels or heads, often surrounded by showy bracts); and habit (shrubs or trees).

Bunchberry

Cornus canadensis

This attractive low subshrub, which is also known as Canada dogwood, spreads by rhizomes, often forming dense populations.

Leaves: Borne in a whorl of four to six at the top of the stem; largest leaves up to about 3 inches (7.5 cm) long.

Flowers: Small, greenish white to purplish, condensed into a head surrounded by four white petal-like bracts.

Fruits: Red "berries" (drupes), single-seeded and edible.

Ecology: Common in moist conifer forests and bogs throughout the Pacific Northwest, from lowlands up to the alpine zone.

Fumitory Family

Fumariaceae

A single species, wild bleedingheart, represents the fumitory family in the Pacific Northwest. These plants typically have bilaterally symmetrical flowers and fernlike leaves.

Wild Bleedingheart

Dicentra formosa

This early flowering perennial herb spreads aggressively by rhizomes. The flowering stems are leafless and stand about 1 foot (30 cm) high.

Leaves: Basal, extensively divided, fern- or parsleylike.

Flowers: Bilaterally symmetrical, pendent at the stem tip, pink to lavender; sepals two; petals four, partially fused, the two largest ones with a saclike base that contains nectar.

Fruits: Many-seeded pods.

Ecology: Common in moist forests and upward into subalpine meadows in the Washington Cascades and adjacent British Columbia.

Bunchberry *Cornus canadensis*

Bunchberry *Cornus canadensis*

Wild Bleedingheart *Dicentra formosa*

EVENING-PRIMROSE FAMILY Onagraceae

The evening-primrose family consists mainly of tropical herbs and shrubs and includes the familiar plant fuchsia. Nonetheless, the family is well represented in the Pacific Northwest, especially by the genus *Epilobium,* the willowherbs (herbs with willowlike leaves). The family is characterized by flowers with parts in fours: four sepals, four petals, eight stamens, and four compartments in the single pistil. The ovary is inferior; that is, borne below the flower parts. The leaves are usually opposite.

Fireweed *Epilobium angustifolium*

Fireweed is aptly named, since it is usually one of the first pioneer plants to invade burned areas. Within the burn, it quickly spreads by seeds and rhizomes, creating a sea of color. An opportunist whose seeds disperse widely, fireweed routinely grows in other disturbed sites, such as roadsides, and frequently inhabits subalpine meadows. This tall, perennial herb may grow to 8 feet (2.5 m)–usually without branching.

Leaves: Alternate, 4 to 8 inches (10 to 20 cm) long, narrowly lance shaped (*angustifolium* means "narrow leaf"), with a conspicuous white midvein.

Flowers: Very showy, petals pink to lavender-purple, up to nearly an inch (2.5 cm) long, borne in elongate, rather dense, and often spirelike racemes.

Fruits: Linear pods or capsules, 2 to 3 inches (5 to 8 cm) long, that produce hundreds of seeds, each with a cottony tuft of hair that facilitates dispersal by the wind.

Ecology: Common throughout the Pacific Northwest, from lowlands to the alpine zone.

A common related species with equally large and showy flowers, **broadleaf willow-herb** *(E. latifolium)* is much shorter than fireweed and has wider leaves (*latifolium* means "broad leaf"). The raceme is much shorter, so the plants have fewer flowers.

Broadleaf willow-herb grows throughout the Pacific Northwest. It is especially common in Alaska and northern British Columbia, where it most frequently inhabits gravel bars in streams and rivers, putting on a beautiful lavender-pink display. In the mountains, it grows on gravelly slopes below snowfields where moisture is available throughout most of the summer.

Fireweed *Epilobium angustifolium*

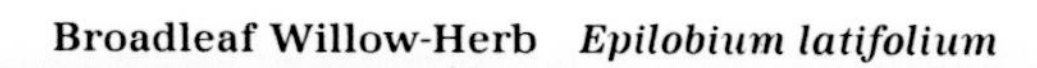

Broadleaf Willow-Herb *Epilobium latifolium*

Yellow Willow-Herb *Epilobium luteum*

Yellow willow-herb, a weak-stemmed perennial herb, may be either erect or sprawling, with stems 1 to 2 feet (30 to 60 cm) long.

Leaves: Opposite, ovate to lance shaped, 1 to 3 inches (2.5 to 7.5 cm) long, toothed.

Flowers: Pale yellow, often drooping; petals about ½ inch (15 mm) long; in general, floral characteristics are typical of the family.

Fruits: Linear pods produce hundreds of tufted seeds.

Ecology: An occasional plant found throughout the Pacific Northwest, growing along streams and in seepage areas from lowlands upward into the subalpine zone.

Alpine Willow-Herb *Epilobium anagallidifolium (alpinum)*

Alpine willow-herb, a rather delicate, erect perennial herb, spreads by short rhizomes, often forming dense mats. The stems grow up to 1 foot (30 cm) high.

Leaves: Opposite, ovate to lance shaped, ⅜ to 1½ inches (1 to 4 cm) long, sometimes toothed.

Flowers: Petals pale pink to rose-purple, about ¼ inch (5 mm) long, notched at the tip (heart shaped); other floral characteristics are typical of the family.

Fruits: Long, linear pods with hundreds of small, tufted seeds.

Ecology: Prefers subalpine meadows and forest openings, often along streams.

Several poorly defined species of *Epilobium* closely resemble alpine willow-herb. In the past some of them have been combined with alpine willow-herb under the binomial *E. alpinum.* All are low, perennial herbs with small, lobed petals that vary in color from white to rose-purple; most of them have erect stems, but some species are strongly mat forming, and all grow in the mountains along streams, on gravelly slopes below snowfields, and in moist meadows. Representatives of this "*alpinum* complex" are distributed throughout the Pacific Northwest. Some of the cushion forms are very attractive, but none can match the color display of fireweed or broadleaf willow-herb.

Yellow Willow-Herb
Epilobium luteum

Alpine Willow-Herb
Epilobium anagallidifolium

Figwort (Snapdragon) Family — Scrophulariaceae

Members of the figwort family have flowers that are bilaterally symmetrical and often highly ornate. The five petals are fused into a tube at the base, with variously designed and positioned lobes that serve to orient the pollinator, usually a bumblebee, as it forages for nectar or pollen. The stamens, usually four, are fused to the petal tube and precisely positioned for effective pollination. The ovary is superior (borne above the other flower parts) and matures into a pod.

Many plants in this family are root parasites that extract water and organic nutrients from their hosts. Among the root parasites are species of *Castilleja* and *Pedicularis.*

Indian Paintbrush — *Castilleja miniata*

The most common paintbrush in western North America, this is a perennial herb with few to several erect, often branched stems arising from a woody root crown. The stems are usually more than 1 foot (30 cm) high, but high-elevation mountain forms are shorter and usually unbranched.

Leaves: Alternate, narrowly lance shaped, usually not lobed, the upper ones bractlike, scarlet colored.

Flowers: Crowded in the axils of the showy bracts; sepals four, fused at the base, scarlet; petals greenish, fused into a narrow base with a long beaklike tip.

Fruits: Many-seeded pods.

Ecology: Grows throughout the Pacific Northwest, from low elevations to subalpine meadows; adapted to a variety of habitats, from dry prairies to stream banks, and from shady forests to open meadows.

Harsh Paintbrush — *Castilleja hispida*

Harsh paintbrush, a perennial, conspicuously hairy herb with erect, clustered, unbranched stems, is 8 to 16 inches (20 to 40 cm) tall.

Leaves: Alternate, divided at the tip into three to five lobes.

Flowers: Similar to Indian paintbrush; sepals and upper bracts red to orange or even yellow.

Fruits: Many-seeded capsules.

Ecology: A common species in Washington and southern British Columbia, growing in forest openings and meadows, from lowlands to the subalpine zone.

Indian Paintbrush *Castilleja miniata*

Harsh Paintbrush *Castilleja hispida*

Harsh Paintbrush *Castilleja hispida*

Small-Flowered Paintbrush *Castilleja parviflora*

A paintbrush common in subalpine meadows, this perennial herb has one or a few erect stems, up to 1 foot (30 cm) high, borne from a thick root crown. Long, silky hair is conspicuous on the upper part of the plant.

Leaves: Alternate, typically divided at the tip into three to five lobes, the central lobe largest.

Flowers: Small (*parvi* means "small"); similar in structure to those of Indian paintbrush; sepals and upper leaves or bracts vary in color from rose-purple to white.

Fruits: Many-seeded capsules.

Ecology: A common resident in subalpine meadows in the region, with a continuous range from southeastern Alaska through Washington.

A number of varieties have been proposed for this species, based primarily on flower color. The white form is variety *albida*, and the rose-purple form is variety *oreopola*.

Alpine Paintbrush *Castilleja rhexifolia*

Alpine paintbrush, an herbaceous perennial, has several clustered stems less than 1 foot (30 cm) high. The plants are unusual because they are hairless or nearly so.

Leaves: Alternate, narrow, and usually not lobed except for the colored bracts at the stem tip.

Flowers: Sepals and upper leaves crimson to scarlet; petals yellowish green, the pointed tip of the fused petals extending slightly above the sepals and bracts.

Fruits: Many-seeded capsule.

Ecology: An alpine plant in the North Cascades of Washington and adjacent British Columbia, growing on rocky substrates.

The paintbrush genus, *Castilleja*, is notoriously difficult taxonomically, and this species is one of the trickiest. Some authors place it only in the Rocky Mountains; others place it in the North Cascades. Because it closely resembles common Indian paintbrush (*C. miniata*) it may be nothing more than an alpine form of that species.

Small-Flowered Paintbrush
Castilleja parviflora var. *albida*

Small-Flowered Paintbrush
Castilleja parviflora var. *oreopola*

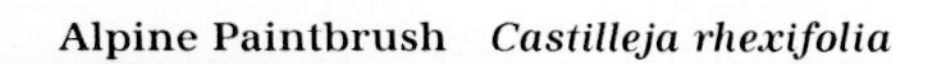

Alpine Paintbrush *Castilleja rhexifolia*

Cliff Paintbrush *Castilleja rupicola*

The brilliant scarlet bracts on cliff paintbrush make it one of the most striking of all mountain wildflowers. And the typical drab colors of its rocky, alpine habitat accentuate its beauty. It is a perennial herb with clustered, low, erect to spreading stems, often cushionlike. The plants are hairy.

Leaves: Alternate but crowded on short stems, deeply divided at the tip into five lobes.

Flowers: Sepals and upper leaves or bracts scarlet; petals green, the beak of the fused petals extending conspicuously beyond the sepals and bracts.

Fruits: Many-seeded capsule.

Ecology: Fairly common in the Washington Cascades, extending northward into southern British Columbia.

Unalaska Paintbrush *Castilleja unalaschcensis*

This perennial herb has lance-shaped leaves and stout, erect, usually clustered stems that may be as much as 2 feet (0.6 m) high. The plants are usually hairy, particularly above.

Leaves: Alternate, narrow (lance shaped), undivided up to the bracts.

Flowers: Sepals and petals greenish yellow, hidden by the bright yellow bracts.

Fruits: Many-seeded capsules.

Ecology: Prefers meadows, but grows in a multitude of habitats from lowlands up to the subalpine zone. Ranges from southeastern Alaska through northern British Columbia.

Less common species of paintbrush grow throughout the Pacific Northwest. Some of them are reddish and others yellow. Among the yellow forms, *C. hyperborea* reaches into Alaska and northern British Columbia. A few technical characteristics separate it from unalaska paintbrush, which it closely resembles. A fairly common scarlet species of subalpine and alpine habitats in the North Cascades of Washington and adjacent British Columbia is **Elmer's paintbrush** (*C. elmeri*). It looks like alpine paintbrush except that it is conspicuously hairy and has divided upper leaves.

Cliff Paintbrush *Castilleja rupicola*

Unalaska Paintbrush
Castilleja unalaschcensis

Elmer's Paintbrush
Castilleja elmeri

Pink Monkeyflower

Mimulus lewisii

This strikingly beautiful wildflower is a rather coarse, perennial herb, up to 3 feet (1 m) tall, with sticky stems and leaves.

Leaves: Opposite, petioles lacking, ovate to lance shaped, 1 to 3 inches (2 to 8 cm) long, usually toothed along the margins, veins very prominent.

Flowers: Bilaterally symmetrical and showy, up to 2 inches (5 cm) long; petals pink-purple, fused into a five-lobed tube, the throat of the tube and the lower petal lobes hairy and marked with yellow; sepals green and fused.

Fruits: Capsules, with thousands of tiny seeds.

Ecology: Locally abundant in the mountains of Washington and southern British Columbia in seepage areas and, especially, along streams in open forests and subalpine meadows.

Mountain Monkeyflower

Mimulus tilingii

This monkeyflower is a low, perennial herb that spreads by shallow rhizomes, often forming extensive mats.

Leaves: Opposite, ovate, up to 1 inch (2.5 cm) long, toothed along the margins; lower leaves stalked, the upper ones much smaller, becoming bractlike; veins conspicuous.

Flowers: Bilaterally symmetrical, brilliant yellow and very showy, about 1 inch (2.5 cm) long; the lower "lip" of the fused petals sculpted with two ridges that bear sticky hairs and red splotches, providing a well-marked landing platform for bumblebees.

Fruits: Capsules, with thousands of tiny seeds.

Ecology: A fairly common species of the Washington Cascades, extending northward into British Columbia; becomes established in seepage areas and along small streams in alpine and subalpine zones.

A similar but much larger plant, **common monkeyflower** *(M. guttatus),* grows predominantly at low elevations. It has succulent but erect stems that bear several flowers. In color, shape, and size the flowers resemble those of the mountain monkeyflower. It, too, prefers wet habitats.

Pink Monkeyflower habitat *Mimulus lewisii*

Pink Monkeyflower *Mimulus lewisii*

Mountain Monkeyflower *Mimulus tilingii*

Bracted Lousewort *Pedicularis bracteosa*

The most common and variable of several regional louseworts, this perennial herb has erect, unbranched stems that stand 2 to 3 feet (0.6 to 0.9 m) tall.

Leaves: Alternate and basal, pinnately compound, narrow leaflets 1 to 3 inches (2.5 to 7.5 cm) long and further divided or deeply toothed, upper leaves becoming bractlike.

Flowers: Bilaterally symmetrical, borne in a dense, bracted spike; petals yellow to occasionally white or even reddish, fused into a tube with an upper hooded lobe, which sometimes has a beaked tip, and a lower "lip" of three smaller lobes; sepals fused, with five teeth; stamens four, fused to the petal tube.

Fruits: Many-seeded capsules.

Ecology: Ranges from southern British Columbia through Washington in moist open forests and, particularly, in subalpine meadows.

Farther north, in British Columbia and Alaska, the similar but short-stemmed **capitate lousewort** *(P. capitata)* grows in meadow habitats. The flowers of this Arctic-alpine species are larger than those of bracted lousewort, but are similar in structure and color and are borne in a dense, terminal head.

Ram's-Horn Pedicularis *Pedicularis racemosa*

A common mountain wildflower, this perennial herb has clustered, erect-to-spreading stems about 1 foot (30 cm) high.

Leaves: Alternate, lance shaped, merely toothed rather than deeply divided.

Flowers: Bilaterally symmetrical, petals white to reddish, fused into a tube that has an upper lobe shaped like a ram's horn or sickle, and a three-lobed lower platform; sepals fused; stamens fused to the petals.

Fruits: Many-seeded capsules.

Ecology: Common in southern British Columbia and Washington in moist open forests or in subalpine meadows, where it usually grows in and around tree clumps.

A somewhat similar alpine and subalpine species, primarily found in the Washington Cascades, is **coiled-beak lousewort** *(P. contorta).* It differs in having mainly basal, pinnately compound leaves with toothed leaflets. The flowers are typically white or cream colored.

Bracted Lousewort *Pedicularis bracteosa*

Capitate Lousewort *Pedicularis capitata*

Ram's-Horn Pedicularis
Pedicularis racemosa

Coiled-Beak Lousewort
Pedicularis contorta

Elephant's Head *Pedicularis groenlandica*

Reddish leaves and stems often characterize this perennial herb, with clumped, erect, unbranched stems up to 1 foot (30 cm) high.

Leaves: Alternate and basal, the blade lance shaped and pinnately divided, with toothed leaflets.

Flowers: Highly ornate, resembling a lavender elephant's head: the upper petal lobe forms the head and trunk, the lower three lobes form the ears and lip, stamens are positioned inside the "mouth"; sepals fused into a tube with five teeth.

Fruits: Many-seeded capsules.

Ecology: Common in wet mountain meadows of southern British Columbia and Washington, ranging from mid-elevations upward to the alpine zone.

A similar species, **bird's beak pedicularis** *(P. ornithorhyncha),* is common in the South Cascades of Washington and infrequent northward to the Coast Mountains of British Columbia. The leaves of this species are mainly basal, and the upper petal lobe resembles a bird's head with a long, curved beak. It grows in moist alpine and subalpine meadows.

These two species in particular and *Pedicularis* in general exemplify the high degree of floral specialization associated with bumblebee pollination. Bees find the hidden rewards, either pollen or nectar, by manipulating the flowers, which requires a combination of skill and strength. The shape of the flower determines the forage behavior of the bees–how the bees enter the flower. The positioning of the stamens and stigma is also correlated with bumblebee behavior.

Sudeten Lousewort *Pedicularis sudetica*

A plant of the Arctic tundra, this perennial herb has erect, mainly solitary, hairy stems up to 1½ feet (45 cm) high.

Leaves: Mainly basal, pinnately compound with divided leaflets.

Flowers: Bilaterally symmetrical, lavender, clustered at the stem tip; upper petal lobe coils downward, and the lower three lobes form a platform with dark maroon markings that guide bees to the nectar.

Fruits: Many-seeded capsules.

Ecology: Primarily an Arctic tundra species, ranging southward into northern British Columbia, where it inhabits moist alpine and subalpine meadows.

lephant's Head *Pedicularis groenlandica*

Elephant's Head *Pedicularis groenlandica*

Bird's Beak Pedicularis
Pedicularis ornithorhyncha

Sudeten Lousewort
Pedicularis sudetica

Davidson's Penstemon *Penstemon davidsonii*

This dwarf shrub is a strikingly beautiful cushion plant. It has spreading, woody branches that root at the nodes, forming dense mats.

Leaves: Evergreen, opposite, small–about ½ inch (12 mm) long and half as wide–leathery and shiny, and usually toothed.

Flowers: Bilaterally symmetrical, 1 inch (2 to 3 cm) long, blue to purple-violet; petals fused into a tube with two upper and three lower lobes (two-lipped); fertile stamens four; anthers densely woolly; a sterile bearded stamen is positioned in the throat of the flower (thus the alternate common name, beard-tongue); sepals five, fused at the base.

Fruits: Many-seeded capsules.

Ecology: Inhabits rocky slopes and ledges in the alpine zone of the Cascades in Washington and adjacent British Columbia.

Small-Flowered Penstemon *Penstemon procerus*

Small-flowered penstemon is a perennial herb with clustered, erect stems 6 to 12 inches (15 to 30 cm) high.

Leaves: Opposite, lance shaped, nontoothed; lower leaves 2 to 4 inches (5 to 10 cm) long, reduced in size upward on the stem, becoming bractlike.

Flowers: Borne in two to three whorls in the axils of the upper leaves or bracts; bilaterally symmetrical, with two upper and three lower lobes, blue to purplish, about ½ inch (8 to 12 mm) long, narrow; anthers four, nonhairy; sterile stamen usually bearded; sepals fused, with five pointed lobes.

Fruits: Many-seeded capsules.

Ecology: Highly variable and grows in a wide variety of habitats, from the sagebrush steppe to rocky alpine slopes. Widely distributed throughout the Pacific Northwest.

A third common penstemon in the Cascades of Washington and adjacent British Columbia is **Cascade penstemon** *(P. serrulatus).* This species differs from the other two in that the stems are clustered and erect or spreading; the leaves are 1 to 3 inches (3 to 8 cm) long, ovate to lance shaped, and sharply toothed; and the showy, blue to purple flowers are about 1 inch (2.5 cm) long. The plants frequent moist, rocky slopes in open forests and subalpine meadows.

Davidson's Penstemon *Penstemon davidsonii*

Small-Flowered Penstemon *Penstemon procerus*

Cascade Penstemon *Penstemon serrulatus*

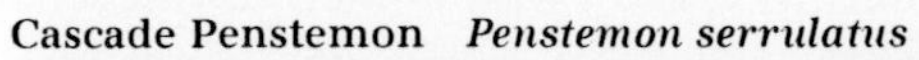

Cusick's Speedwell *Veronica cusickii*

Small and attractive, this is a delicate, perennial herb. The stems are usually clustered and erect, 4 to 8 inches (10 to 20 cm) high. The plant spreads by rhizomes.

Leaves: Opposite, ovate, up to 1 inch (2.5 cm) long and half as wide, nontoothed.

Flowers: Borne in a dense or open raceme at the stem tip; petals four, about ⅜ to ¼ inch (4 to 6 mm) long, the lower the smallest, and the upper the largest; petals slightly fused at the base, blue-violet; sepals four; stamens two, fused to the petals.

Fruits: Heart-shaped capsules with few to several seeds.

Ecology: This plant prefers moist to wet meadows and rocky slopes, often growing along streams in the subalpine and, occasionally, alpine zones; common in the high mountains of Washington and rare northward into southern British Columbia.

Alpine Speedwell *Veronica wormskjoldii*

The most common species of *Veronica,* alpine speedwell is a delicate, rhizomatous perennial, with clustered, erect stems 2 to 8 inches (5 to 20 cm) high.

Leaves: Mainly opposite (sometimes alternate above), ovate to elliptical, about 1 inch (2 to 4 cm) long, sometimes with small teeth.

Flowers: Few to several in tight racemes or spikes at the stem tip; the flowers are similar to those of Cusick's speedwell but smaller, the petals only about ⅛ inch (3 to 4 mm) long and generally paler blue.

Fruits: Heart-shaped capsules with numerous tiny seeds.

Ecology: Ranges throughout the mountains of the Pacific Northwest in moist to wet areas, especially along stream banks, in the subalpine and alpine zones.

The high mountains of the Olympic Peninsula are home to a much more attractive and somewhat similar species, **cut-leaf synthyris** *(Synthyris pinnatifida).* Like the speedwells, its flowers are somewhat bilaterally symmetrical and have four blue, partly fused petals, four sepals, two stamens, and a heart-shaped capsule. The cut-leaf synthyris leaves, however, are pinnately divided, fernlike, and all basal, not at all like those of the speedwells. Also, "wool" covers the stems.

Cusick's Speedwell *Veronica cusickii*

Alpine Speedwell *Veronica wormskjoldii*

Cut-Leaf Synthyris *Synthyris pinnatifida*

Gentian Family — Gentianaceae

The small gentian family comprises annual and perennial herbs that have opposite, somewhat succulent leaves. The flowers are showy and borne singly or in clusters at the tip of erect stems. The petals are fused, becoming tubular or bell shaped, with four or five lobes. Plaits or scales often occur between the petal lobes. The sepals are fused together, and the stamens are fused to and derived from the petal tube.

Mountain Bog Gentian — *Gentiana calycosa*

Mountain bog gentian has clustered, unbranched stems up to 1 foot (30 cm) high. The plants are hairless and more or less shiny.

Leaves: Opposite, crowded on the stem, ovate, without petioles, up to 1 inch (2.5 cm) long.

Flowers: Showy, 1 inch (2.5 cm) long or more, borne singly or few at the stem tip; petals dark blue, fused into a broad five-lobed tube with fringed plaits between the lobes.

Fruits: Many-seeded pods (capsules).

Ecology: A frequent inhabitant of wet meadows and bogs in the Cascade and Olympic Mountains of Washington, ranging south to California and east into the Rocky Mountains.

Glaucous Gentian — *Gentiana glauca*

This is a small perennial herb with unbranched stems not more than 6 inches (15 cm) long. Plants spread by rhizomes.

Leaves: Mainly in a basal rosette, stem leaves opposite, in two to four pairs, small and succulent.

Flowers: About ½ inch (12 mm) long, clustered at the stem tip; petals dark blue to greenish, fused into a narrow tube with five lobes and small plaits between the lobes.

Fruits: Many-seeded pods.

Ecology: A common plant of alpine and subalpine meadows in British Columbia and Alaska, barely reaching the North Cascades of Washington.

In Alaska and northern and coastal British Columbia, **Douglas gentian** *(G. douglasii)* is a frequent inhabitant of bogs and wet meadows. It differs from *G. glauca* in having white petals with spreading lobes and a more open flower cluster.

Mountain Bog Gentian *Gentiana calycosa*

Glaucous Gentian *Gentiana glauca*

Douglas Gentian *Gentiana douglasii*

Geranium Family

Geraniaceae

In the Pacific Northwest, members of the small geranium family can be easily identified by a combination of floral and leaf characteristics. The flower parts of these perennial herbs are in multiples of five. The pistils are important diagnostically, because they develop a long "beak" as they mature, which accounts for common names like crane's bill and stork's bill. The leaves are palmately divided into three to five major segments. Most species have a geranium aroma.

Northern Geranium

Geranium erianthum

Northern geranium is one of the many plants that spreads by rhizomes. The stems are hairy, usually branched, and 1 to 2 feet (30 to 60 cm) tall.

Leaves: Palmately divided into three to five segments, which are further divided and toothed; basal leaves with long stalks, stem leaves stalkless.

Flowers: Lavender-purple, petals ½ to 1 inch (1.5 to 2.5 cm) long.

Fruits: The beaked pistil divides into five, one-seeded "nutlets," each with a long, coiled style that may aid in seed dispersal.

Ecology: A plant of moist, open forests or meadows, from lowlands up to the subalpine zone; fairly common in northern British Columbia and southeastern Alaska.

Sticky Geranium

Geranium viscosissimum

This attractive flowering plant has somewhat succulent branched stems, which may be erect or sprawling and grow to about 2 feet (60 cm) tall. The entire plant is sticky, as indicated by both the common and Latin names (*viscosissimum* means "sticky").

Leaves: Palmately divided and generally similar to those of northern geranium but sticky from glandular hairs.

Flowers: Showy, petals pink-purple with darker veins, up to 1 inch (2.5 cm) long.

Fruits: Similar to those of northern geranium, but sticky.

Ecology: Open forests to subalpine meadows, mainly on the eastern side of the Cascades in Washington and adjacent British Columbia.

A third species, **Richardson's geranium** *(G. richardsonii),* grows in the Pacific Northwest, overlapping in distribution with the other two species. White to pale pink flowers characterize Richardson's geranium.

Northern Geranium *Geranium erianthum*

Sticky Geranium *Geranium viscosissimum*

Richardson's Geranium *Geranium richardsonii*

HAREBELL FAMILY — Campanulaceae

Campanula means "bell shaped," which describes the attractive blue flowers in this family. The flowers may be either radially or bilaterally symmetrical, depending on the species. The petals are fused into a tube with five lobes, or teeth. The flowers have five sepals and five stamens, and the ovary is inferior, situated below the other flower parts.

Common Harebell — *Campanula rotundifolia*

Common harebell is a variable perennial herb. The stems range in height from more than 2 feet (0.6 m) in low forest forms to a few inches in alpine forms.

Leaves: Basal leaves roundish with long stalks; stem leaves linear, not toothed, alternate and/or opposite.

Flowers: Radially symmetrical, bell shaped, and nodding or erect, sky blue, about 1 inch (2.5 cm) long; fused petals form the "bell" with five teeth.

Fruits: Many-seeded capsules.

Ecology: More or less common throughout the mountains of the Pacific Northwest in a variety of habitats, from low forests to alpine ridges.

Common harebell is an extremely variable species. At low elevations the plants are tall and bear several flowers on branched stems. The alpine forms are low and usually bear a single, larger flower.

The alpine form of common harebell closely resembles **mountain harebell** *(C. lasiocarpa).* This is a rather common and attractive plant of the Alaskan tundra and the mountains of British Columbia, where it is found mainly on rocky subalpine or alpine slopes. It differs from the common harebell by having somewhat larger flowers with hairy sepals and distinctively toothed leaves.

Common Harebell *Campanula rotundifolia*

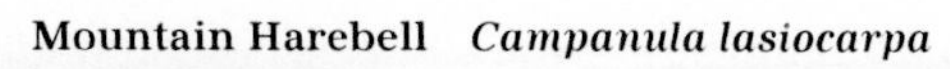

Mountain Harebell *Campanula lasiocarpa*

HEATH FAMILY

Ericaceae

Ecologically, the heath family is one of the most important flowering plant families in the Pacific Northwest, even though it has far fewer species than other families, such as Asteraceae. Almost every Northwest habitat has some heath family species, and sometimes they are dominant members of the community. They also provide abundant food and shelter for wildlife.

The family is diverse and includes trees, shrubs, herbs, and even nongreen saprophytes. The species are linked, although loosely, by flower form and structure. The flowers parts are mostly in fives: five sepals, five petals, ten stamens, and five compartments in the ovary. In most heath species, the petals are fused into an urn-shaped or bell-shaped structure. All species have the unusual characteristic of the anthers shedding their pollen through terminal pores rather than longitudinal slits. In many species, especially those with pendant flowers, the anthers have elaborate antlerlike appendages called awns. The awns impede entrance into the urn-shaped petal tube, causing visiting insects, especially bumblebees, to rattle the anthers as they probe a flower, shaking pollen from the terminal pores onto their bodies–nature's saltshaker.

Indian-Pipe

Monotropa uniflora

A nongreen (white), succulent herb with waxy, clustered stems, Indian pipe eventually reaches up to 1 foot (30 cm) high. The common name comes from the solitary, nodding flowers.

Leaves: Small, white, and scalelike, alternate.

Flowers: Solitary (*uniflora*) at the stem tip, white, initially nodding and bell shaped, later becoming erect.

Fruits: Erect, many-seeded capsules.

Ecology: Understory species in conifer forests, from low to mid-elevations; common on the western side of the Cascades in Washington and adjacent British Columbia.

Indian-pipe and other nongreen ericads depend on associated fungi for their nutritional needs. This trophic specialization enables the plants to grow in deep shade and avoid competition from other plants.

Another interesting nongreen ericad with a similar range is **candystick** *(Allotropa virgata).* It has 6- to 18-inch (15 to 45 cm) stems, with alternating pink and white stripes and scalelike, pinkish brown leaves. The flowers, which occur in dense, elongate spikes, are small, lack petals, and have conspicuous red anthers.

Pinesap *(Hypopitys monotropa)* resembles Indian-pipe, as the shared Latin name *monotropa* suggests, but has clustered flowers and a general coloration varying from pale yellow to reddish. Although similar in ecological requirements, it ranges to northern British Columbia.

Indian-Pipe *Monotropa uniflora*

Candystick *Allotropa virgata*

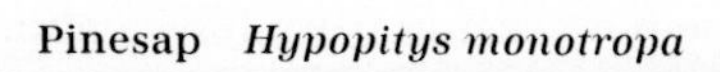

Pinesap *Hypopitys monotropa*

Pinedrops *Pterospora andromedea*

An unusual plant, pinedrops is a succulent perennial herb with single to few unbranched stems up to 3 feet (1 m) high, the entire plant reddish brown (nongreen) and conspicuously sticky. Like other nongreen ericads, pinedrops depends on associated fungi for nutrition.

Leaves: Scalelike, reddish brown, crowded toward the stem base.

Flowers: Small, about ¼ inch (6 mm) long, yellowish brown, round to urn shaped, pendant on short stalks; borne in a dense, elongate raceme.

Fruits: Round capsules.

Ecology: This plant follows the Rocky Mountains from Alaska to Mexico; in the Pacific Northwest, it grows on the drier, eastern side of the Cascades, especially in pine forests.

Kinnikinnick (Bearberry) *Arctostaphylos uva-ursi*

Kinnikinnick, also called bearberry, has a long history of commercial use. It is a highly adaptable ground cover, and the leaves have been dried and used commercially as a tobacco substitute. More important, however, the species has considerable ecological value. The berries are edible and persist on the branches through winter, providing a food source during a time of scarcity. The flowers appear early in spring and serve as an important source of nectar for queen bumblebees emerging from hibernation. Kinnikinnick plants are dominant in their habitat and greatly influence their immediate environment.

An attractive evergreen shrub with prostrate branches that root at the nodes, kinnikinnick forms dense, expansive mats. The bark is reddish.

Leaves: Alternate, ovate, broader toward the tip, about 1 inch (2.5 cm) long, evergreen, leathery, and shiny.

Flowers: Urn shaped and pendant, white to more often pink, clustered at the tips of branches.

Fruits: Bright red, pulpy berries.

Ecology: Ranges from the coast to alpine ridges, growing on rocky, exposed sites in suitable habitats throughout the Pacific Northwest.

A related but hairier species of kinnikinnick, ***A. nevadensis,*** grows in the foothills on the eastern side of the Washington Cascades. The two species apparently hybridize where their ranges overlap.

Pinedrops *Pterospora andromedea*

Kinnikinnick fruits *Arctostaphylos uva-ursi*

Kinnikinnick flowers *Arctostaphylos uva-ursi*

Red Bearberry

Arctostaphylos alpina var. *rubra*

Dwarfed and prostrate, red bearberry is a common shrub with deciduous leaves. The bark is thin and more or less papery. It is attractive, especially in autumn when the leaves turn brilliant red, matching the color of the berries. The berries are edible but not flavorful.

Leaves: Alternate, ovate, about 1 inch (2.5 cm) long, somewhat leathery; upper surface conspicuously marked with a network of sunken veins; leaf margins toothed.

Flowers: Urn shaped, nodding, white to pink, clustered at the tip of narrow branches.

Fruits: Bright red, juicy berries.

Ecology: Common in the Alaskan tundra and open spruce forests, extending southward into northern British Columbia; inhabits lowland tundra and bogs upward into moist subalpine and alpine meadows.

A closely related shrub of the Arctic and alpine tundra is the **alpine bearberry** *(A. alpina* var. *alpina).* The leaves of the two varieties are similar in size and shape, and both have conspicuous vein networks. Both produce edible berries, but alpine bearberries are usually purplish black. Alpine bearberry is common in Alaska and the Canadian Rockies, barely reaching into northern British Columbia.

Bog Rosemary

Andromeda polifolia

An enchanting dwarf evergreen shrub, this bog species has erect to spreading branches up to 2 feet (60 cm) long.

Leaves: Alternate, narrowly elliptical to linear, sharp pointed, about 1 inch (2 to 3 cm) long, leathery and persistent, whitish on the undersurface and grooved on top.

Flowers: Urn shaped and small, about ¼ inch (5 to 8 mm) long, pinkish, borne on long stalks in clusters of two to several at the branch tips.

Fruits: Small, round capsules.

Ecology: Common in the Alaskan tundra, ranging south to southern British Columbia, where it grows from low to mid-elevations in the mountains.

Red Bearberry *Arctostaphylos alpina* var. *rubra*

Alpine Bearberry *Arctostaphylos alpina* var. *alpina*

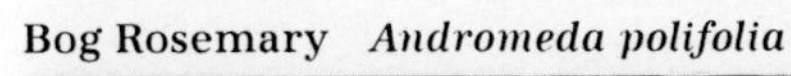

Bog Rosemary *Andromeda polifolia*

White Mountain Heather *Cassiope mertensiana*

The branches of white mountain heather resemble clubmosses, and the two often grow together. The close resemblance is responsible for the frequently used common name, moss heather. This is a low, often matted, evergreen shrub, 8 to 12 inches (20 to 30 cm) high, with spreading branches.

Leaves: Opposite, narrowly ovate and scalelike, about ¼ inch (6 mm) long, crowded, completely concealing the stem, occurring in four rows (two sets of opposite leaves).

Flowers: Looking like delicate white bells, about ¼ inch (5 to 8 mm) long; sepals red, contrasting with the fused, white petals; flowers borne on stalks near the branch tips.

Fruits: Small, woody capsules.

Ecology: A widespread shrub throughout the Pacific Northwest, preferring well-drained but moist soils on subalpine and, occasionally, alpine slopes.

A similar but far less common heather is **four-angled mountain heather** *(C. tetragona).* It differs from white mountain heather in that the leaves are grooved on the undersurface, and the branches are more conspicuously four-angled. Also, the sepals are smaller and yellow-orange rather than red. Four-angled mountain heather is common in the Alaskan tundra, extending southward along mountain ranges, where it is predominantly an alpine species. It is uncommon in southern British Columbia and rare in Washington.

Alaskan Mountain Heather *Cassiope stelleriana*

Alaskan mountain heather is a dwarf, spreading, often matted, evergreen shrub, with stems 4 to 6 inches (10 to 15 cm) high.

Leaves: Alternate, lance shaped to needlelike, about ⅛ to ¼ inch (3 to 5 mm) long, projected outward on the branches.

Flowers: White, bell shaped, and nodding, usually solitary at branch tips; sepals reddish.

Fruits: Small, woody capsules.

Ecology: As the common name suggests, this plant is widespread in Alaska. Extends southward along the western ranges of British Columbia into the Washington Cascades, preferring rocky but moist alpine habitats such as seepage areas.

When the plants are not in flower, they can be easily confused with alpine azalea, crowberry, and heathers in the genus *Phyllodoce.*

White Mountain Heather *Cassiope mertensiana*

Four-Angled Mountain Heather *Cassiope tetragona*

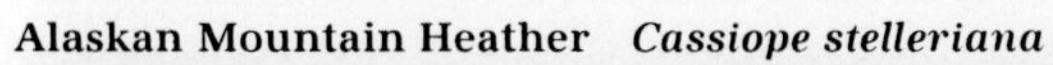

Alaskan Mountain Heather *Cassiope stelleriana*

Pink Mountain Heather *Phyllodoce empetriformis*

An attractive heather, this species is a low, extensively branched, and spreading evergreen shrub. The plants are 6 to 18 inches (15 to 45 cm) high.

Leaves: Alternate, closely crowded, linear, and needlelike, up to ½ inch (12 mm) long, grooved on the undersurface.

Flowers: Bell shaped, nodding to erect, pale pink to dark rose, borne in showy clusters of a few to several at the branch tips.

Fruits: Woody capsules.

Ecology: Distributed throughout the Pacific Northwest, growing on moist but well-drained slopes, frequently in open, high-elevation forests but most commonly in subalpine meadows and protected sites with deep soil in the alpine zone.

As the Latin name *(empetriformis)* suggests, this species closely resembles crowberry *(Empetrum nigrum)*. Crowberry, however, has inconspicuous flowers and a black berry for a fruit.

Yellow Mountain Heather *Phyllodoce glanduliflora*

This heather, low and extensively branched and spreading, is an evergreen shrub that often forms extensive mats. The stems are 6 to 18 inches (15 to 45 cm) long.

Leaves: Alternate, crowded on the stem, linear, and needlelike, up to ½ inch (12 mm) long, covered by glandular hairs.

Flowers: Narrowly bell shaped to urn shaped, dull yellow; the fused petals, sepals, and flower stalk are covered by glandular hairs, thus the Latin name *glanduliflora*.

Fruits: Woody capsules.

Ecology: Distributed throughout the Pacific Northwest and especially common near the treeline of the subalpine zone and upward into the alpine zone. Prefers moist, well-developed soils.

The range of this species frequently overlaps with that of pink mountain heather, resulting in hybridization. The hybrids, which share characteristics of both parental species, have been referred to as *Phyllodoce intermedia*.

Aleutian heather *(P. aleutica)* resembles yellow mountain heather, and the two species are sometimes placed in the same species. Both have glandular flowers, but in Aleutian heather the flowers are broader, shorter, and cream colored rather than dull yellow. Aleutian heather grows in moist meadows and bogs, barely reaching into southeastern Alaska from the Aleutians.

Pink Mountain Heather *Phyllodoce empetriformis*

Yellow Mountain Heather *Phyllodoce glanduliflora*

Aleutian Heather *Phyllodoce aleutica*

Alpine Azalea *Loiseleuria procumbens*

A plant of the northern reaches, this dwarf evergreen shrub forms very dense mats.

Leaves: Opposite, elliptical, small, about ¼ inch (4 to 8 mm) long, persistent, leathery, shiny, grooved on the upper surface.

Flowers: Small, pale to dark pink, the petal lobes spreading; borne in the axils of upper leaves.

Fruits: Small, woody capsules.

Ecology: Common in the Alaskan tundra and the alpine zone of northwestern British Columbia, extending southward to northern Washington, where it is a rare species.

False Azalea *Menziesia ferruginea*

Also known as fool's huckleberry, this is a variable, deciduous shrub 3 to 9 feet (1 to 3 m) tall.

Leaves: Alternate but clustered at branch tips, ovate to elliptical, about 2 inches (4 to 6 cm) long, margins usually ragged or toothed.

Flowers: Urn shaped with only four teeth or lobes, about ¼ inch (6 to 8 mm) long, yellowish to copper or rust colored (ferruginous); flowers clustered on glandular stalks at stem tips.

Fruits: Woody capsules.

Ecology: A plant of moist conifer forests from low to high elevations, often growing in bogs and along streams; common in Washington and southern British Columbia.

Woodnymph *Moneses uniflora*

This plant is also called single delight, a translation of *moneses* that relates to the single, attractive, delightfully fragrant flower. Woodnymph is a delicate subshrub with woody lower stems and evergreen leaves. It spreads by woody rhizomes.

Leaves: Mainly basal, blades ovate to circular, up to 1 inch (2.5 cm) long, leathery and shiny, conspicuously veined and toothed.

Flowers: White, about 1 inch (2.5 cm) across, petals spreading and usually ragged along the margins; flowers solitary and facing downward at the top of a stem 2 to 4 inches (5 to 10 cm) long.

Fruits: Capsules.

Ecology: Grows in deep humus and wet, rotting wood in moist forests throughout the Pacific Northwest at low to mid-elevations; rarely common.

Alpine Azalea *Loiseleuria procumbens*

False Azalea *Menziesia ferruginea*

Woodnymph *Moneses uniflora*

Slender Wintergreen *Gaultheria ovatifolia*

This dwarf, evergreen shrub has trailing branches that root at the nodes.

Leaves: Alternate, leathery and shiny, evergreen, toothed, ovate, 1 to 2 inches (2 to 5 cm) long.

Flowers: Bell shaped, white to pale pink, about ⅛ inch (3 to 5 mm) long, borne on the underside of branches in leaf axils; sepals hairy.

Fruits: Red berries.

Ecology: Locally common plant in forests and on open slopes in the mountains of Washington, ranging southward into Oregon and northward into southern British Columbia.

A related shrub with smaller leaves and flowers is **alpine wintergreen** *(G. humifusa).* It also differs from slender wintergreen in that the sepals are hairless, and it requires a moister habitat. The distribution of the two species is similar.

Salal *Gaultheria shallon*

Ecologically, salal is an important shrub. Often, it is a major dominant in bogs and forest communities, and its berries are eaten by a variety of animals. It is also a valued horticultural plant, used extensively in landscaping and flower arrangements. Salal is a low to more often medium-sized evergreen shrub with spreading to erect stems.

Leaves: Alternate, blades ovate to elliptical, 2 to 4 inches (5 to 10 cm) long and half as wide, toothed, leathery, shiny, and conspicuously veined.

Flowers: Urn shaped, white to pink, ¼ to ⅜ inch (6 to 10 mm) long, borne with bracts in elongate racemes in leaf axils near the branch tips; the flowers are covered with glandular hairs.

Fruits: Purple to black berries; the fleshiness is the product of the five glandular, pulpy sepals.

Ecology: Common in a variety of habitats, from bogs to dry, well-drained slopes. Most abundant in dry Douglas fir forests, from the coast to mid-elevations on the western side of the Cascades and Olympics. Ranges north into the coastal mountains of southern British Columbia.

Slender Wintergreen *Gaultheria ovatifolia*

Salal fruits *Gaultheria shallon*

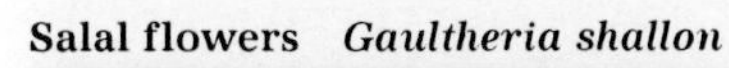

Salal flowers *Gaultheria shallon*

Bog Laurel

Kalmia microphylla

A wide-ranging species, bog laurel is a low-growing evergreen shrub, with spreading branches that root at the nodes. Bog laurel flowers have an interesting pollination strategy. The stamens lie in pockets and are "spring-loaded." When an insect visitor touches them, they spring out forcefully, showering the insect with pollen.

Leaves: Opposite, elliptical, persistent and leathery, about 1 inch (2.5 cm) long, dark green above and gray beneath.

Flowers: Showy, light to dark pink, about ½ inch (10 to 14 mm) across, clustered on long stalks at the tips of erect branches; petals fused and saucer shaped with five lobes and ten "pockets" in which the stamens initially lie.

Fruits: Woody capsules.

Ecology: A common inhabitant of wet, acidic mountain meadows and bogs; distributed more or less throughout the Pacific Northwest.

This delightful little shrub shows great variation in size. A more robust, lowland bog form **western bog laurel** *(K. microphylla* ssp. *occidentalis),* is often placed in a separate variety or subspecies. However, the variation appears continuous, with intergradation at intermediate elevations.

Labrador Tea

Ledum groenlandicum

Resins give this plant a distinctive aroma and taste, resulting in its popularity for tea. It is an erect, evergreen shrub, up to 6 feet (2 m) tall.

Leaves: Alternate to opposite, narrowly elliptical, 1 to 2 inches (2 to 5 cm) long; the upper side of the leaf blade is strongly veined and deeply grooved; the underside has the edges rolled under and is densely covered with rust-colored hairs.

Flowers: White, showy, about ⅜ inch (1 cm) across, borne in large umbrella-shaped clusters at the branch tips; petals five, spreading but fused at the base.

Fruits: Woody capsules.

Ecology: This is a "bog indicator" species, restricted to bogs and highly acidic soil at low to mid-elevations throughout the Pacific Northwest.

A related species growing at higher elevations in conifer forests of Washington and southern British Columbia is **trapper's tea** *(L. glandulosum).* Its leaves are broader and the undersurface is covered with glands, rather than rusty hairs.

Bog Laurel *Kalmia microphylla*

Labrador Tea *Ledum groenlandicum*

Trapper's Tea *Ledum glandulosum*

Prince's Pine *Chimaphila umbellata*

Also referred to as pipsissewa, this is a low, erect, evergreen shrub, usually less than 1 foot (30 cm) high. It spreads by rhizomes.

Leaves: Mainly whorled, in one to two whorls, narrowly elliptical, sharply toothed, persistent, leathery, and shiny.

Flowers: Pinkish, about ½ inch (10 to 14 mm) across, facing downward, three to several borne in umbrella-like clusters at the stem tip; petals five, spreading and not fused.

Fruits: Capsules.

Ecology: A plant of conifer forests from low to mid-elevations throughout the Pacific Northwest, most commonly in Washington and southern British Columbia.

This princely, miniature "pine" overlaps in distribution in Washington and southern British Columbia with **little pipsissewa** *(C. menziesii),* a close relative with somewhat smaller leaves and only two or three flowers per stem. Also, the flowers are lighter pink or even white.

Pink Wintergreen *Pyrola asarifolia*

Wintergreen, the common name for this and other *Pyrola* species, relates to the leaves persisting through one winter. At any time, the plant exhibits a mixture of leaves from the current year and the previous year. The result is an increase in the effective length of the plant's photosynthetic season. Pink wintergreen is a perennial, erect herb, 6 to 15 inches (15 to 37 cm) high. It spreads by slender rhizomes.

Leaves: Basal, the blades ovate to heart shaped or circular, 1 to 3 inches (2.5 to 7.5 cm) long and as wide, sometimes toothed, dark green, and shiny.

Flowers: Pinkish to purplish, about ½ inch (10 to 15 mm) wide, generally facing downward, numerous in an elongate, narrow raceme; style elongate and bent.

Fruits: Capsules.

Ecology: Wide-ranging throughout the region in moist conifer forests from low to high elevations; prefers humus-rich acidic soils and bogs.

Pink wintergreen is the showiest of several *Pyrola* species that grow in the Pacific Northwest. It is distinguished by a combination of broad, nonmottled leaves, pinkish to purple flowers, and the bent style.

Prince's Pine *Chimaphila umbellata*

Pink Wintergreen *Pyrola asarifolia*

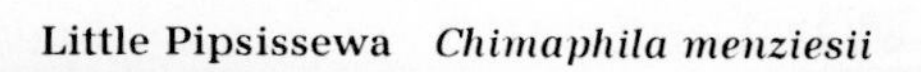

Little Pipsissewa *Chimaphila menziesii*

Green Wintergreen *Pyrola chlorantha*

Because this is an inconspicuous plant with pale flowers and few, featureless leaves, it is probably more common than reports indicate. This erect, perennial herb, 4 to 10 inches (10 to 25 cm) high, spreads by rhizomes.

Leaves: All basal, one or few per flowering stem, ovate to circular, up to 1 inch (2 to 3 cm) long, sometimes toothed.

Flowers: White to greenish yellow, about ½ inch (9 to 12 mm) across, facing downward; the styles extend beyond the petals, and are bent; flowers few to several in a narrow raceme.

Fruits: Capsules.

Ecology: An occasional species that prefers deep, humus-rich soil at low to mid-elevations in the mountains; distributed throughout the Pacific Northwest.

A related species with similarly unspectacular, pale yellow to greenish (rarely purplish) flowers is **white-veined wintergreen** *(P. picta)*. However, its leaves are conspicuous on the forest floor, white veins standing out on the dark green leaf. Also, the flowering stems are usually red, adding to its overall appeal. This wintergreen ranges from southern British Columbia through Washington, mainly on the eastern side of the Cascades.

One-Sided Wintergreen *Orthilia (Pyrola) secunda*

This is the most common wintergreen in the region. It is an evergreen, perennial subshrub (the lower part of the stem is woody), and spreads by rhizomes. The stems are mostly single, 2 to 6 inches (5 to 15 cm) high.

Leaves: Basal and alternate on the lower part of the stem; blades ovate or elliptical to circular, usually toothed, bright green, leathery, and shiny, 1 to 2 inches (2 to 5 cm) long.

Flowers: Petals greenish white; style extended and straight; flowers small, bell shaped, and nodding, concentrated on one side of the flowering stem.

Fruits: Capsules.

Ecology: Extends from lowlands to the subalpine zone in shady, coniferous forests as well as open areas throughout the Pacific Northwest.

Green Wintergreen *Pyrola chlorantha*

One-Sided Wintergreen *Orthilia secunda*

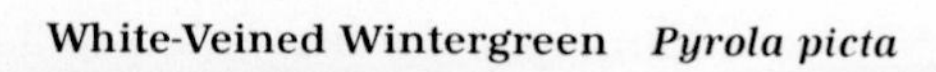

White-Veined Wintergreen *Pyrola picta*

Pacific Rhododendron *Rhododendron macrophyllum*

Although Pacific rhododendron is the state flower of Washington, its distribution in that state is very restricted. Nonetheless, its beauty is unsurpassed and it is frequently grown as a cultivated landscape plant. It is a large, evergreen shrub, up to 12 feet (4 m) tall.

Leaves: Alternate and opposite, ovate to elliptical, large (*macrophyllum* means "large leaf"), 3 to 8 inches (8 to 20 cm) long, persistent, leathery, and nontoothed.

Flowers: Very showy, 1 to 2 inches (2.5 to 5 cm) across, clustered in large numbers at the branch tips; petals pink to purplish, fused, and broadly bell shaped with five wavy lobes; stamens ten, of unequal lengths.

Fruits: Large, woody capsules.

Ecology: Restricted to the western side of coastal mountains from California to southern British Columbia. A common understory dominant species in some of the "temperate rain forests."

White Rhododendron *Rhododendron albiflorum*

The white-flowered rhododendron is a medium-sized, deciduous shrub. It is extensively branched and grows up to 9 feet (3 m) tall.

Leaves: Alternate but often appearing to be whorled, especially at the stem tips, shiny green on top, paler beneath, 2 to 3 inches (4 to 8 cm) long; the leaves often have shallow, rounded teeth.

Flowers: White, about 1 inch (2.5 cm) across, borne in clusters of two to several along the branches; petals fused at the base, forming a broad bell with five spreading lobes.

Fruits: Woody capsules.

Ecology: Prefers moist meadows, especially around tree clumps, in open conifer forests and along mountain streams, from west-central British Columbia through Washington.

A species closely related to these rhododendrons is **copper-bush** *(Cladothamnus pyroliflorus)*, an attractive shrub with distinctive copper-colored flowers that are intermediate in size and shape between those of white rhododendron and some *Pyrola* species. Its leaves are leathery but deciduous, alternate, narrowly elliptical, and 1 to 2 inches (2 to 5 cm) long. It grows in moist conifer forests and subalpine meadows, often along stream banks, ranging from Alaska through Washington, but seldom is it common.

Pacific Rhododendron
Rhododendron macrophyllum
–Diane Doss photo

White Rhododendron
Rhododendron albiflorum

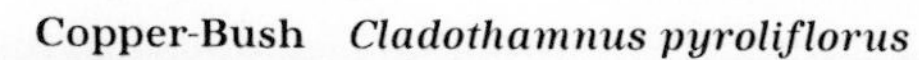

Copper-Bush *Cladothamnus pyroliflorus*

Alaskan Blueberry *Vaccinium alaskaense*

This medium-sized, deciduous shrub grows up to 5 feet (1.5 m) tall. The twigs are somewhat angled and have yellow-green bark.

Leaves: Alternate, deciduous, ovate to elliptical, 1 to 2 inches (2.5 to 6 cm) long, usually nontoothed; the midvein on the undersurface lightly covered with glandular hairs.

Flowers: Broadly urn shaped, with five short petal lobes, about ¼ inch (7 mm) long and at least as wide, pink to bronze colored; sepals fused and saucer shaped with five indistinct lobes.

Fruits: Juicy, dark purple berries, often coated with a pale blue, fine powdery "bloom."

Ecology: One of the most common *Vaccinium* species on the western side of the Washington Cascades and Coast Mountains of British Columbia and Alaska, extending from lowlands to the subalpine zone as a frequent understory dominant species in moist conifer forests.

Oval-Leaved Blueberry *Vaccinium ovalifolium*

Another medium-sized deciduous shrub, oval-leaved blueberry grows up to 4 feet (1.2 m) tall. The twigs are distinctly angled and have yellow-green bark.

Leaves: Alternate, deciduous, ovate to elliptical, 1 to 2 inches (2 to 5 cm) long, usually nontoothed; undersurface does not have glandular hair.

Flowers: Urn shaped, with five short petal lobes, about ¼ inch (7 mm) long, longer than wide, pale to dark pink; sepals fused and saucer shaped, with five indistinct lobes.

Fruits: Juicy, dark purple berries, usually coated with a pale blue, fine powdery "bloom."

Ecology: Distributed more or less throughout the Pacific Northwest in moist conifer forests and meadows, often in bogs, from lowlands into the subalpine zone.

Oval-leaved and Alaskan blueberry plants are difficult to distinguish when they are not in bloom; the flowers of oval-leaved blueberry appear before or with the leaves, but the flowers of Alaskan blueberry appear after the leaves are well developed. Also, Alaskan blueberry flowers are as wide or wider than long; oval-leaved blueberry flowers are longer than wide. The berries from both species are edible and serve as food for a variety of animals, including humans.

Alaskan Blueberry flowers
Vaccinium alaskaense

Oval-Leaved Blueberry
Vaccinium ovalifolium

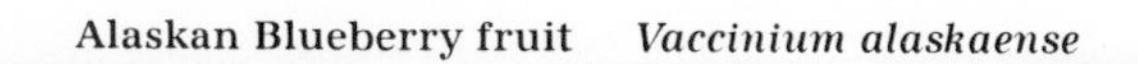

Alaskan Blueberry fruit *Vaccinium alaskaense*

Cascade Huckleberry *Vaccinium deliciosum*

The Latin *deliciosum* aptly describes Cascade huckleberry—the berries are truly delicious. Cascade huckleberry grows in profusion on high mountain slopes, making the fruit especially attractive to berry pickers, including bears and birds. The importance of the plants' sheer numbers becomes obvious in autumn, when the leaves turn red and entire mountain slopes light up in a brilliant display of color. This shrub is low, spreading, deciduous, often matted, and usually less than 1 foot (30 cm) high.

Leaves: Alternate, deciduous, ovate to elliptical, ½ to 2 inches (1.5 to 5 cm) long and about half as wide, sometimes toothed toward the tip.

Flowers: Broadly urn shaped, with five short petal lobes, about ¼ inch (6 to 7 mm) long and nearly as wide, pinkish; sepals fused and saucerlike, with five indistinct lobes.

Fruits: Juicy, blue berries.

Ecology: A conspicuous dominant species in subalpine meadows and on alpine ridges throughout the Cascades and Olympics of Washington, extending into southern British Columbia.

A similar high-mountain species, **dwarf huckleberry** *(V. caespitosum),* differs from Cascade huckleberry by having flowers about twice as long as wide and leaves more distinctly toothed. It is common at northern latitudes.

Mountain Huckleberry *Vaccinium membranaceum*

As the name implies, this is a high-elevation shrub, medium sized and deciduous, growing up to 6 feet (2 m) tall.

Leaves: Alternate, ovate to elliptical, sharp pointed, 1 to 3 inches (2.5 to 7 cm) long, conspicuously veined and toothed.

Flowers: Urn shaped, with five petal lobes, about ¼ inch (6 mm) long, longer than wide, pink to yellowish; sepal lobes indistinct.

Fruits: Dark purple to black, rarely red, rather acidic but with excellent flavor.

Ecology: A common shrub in high-elevation forests and forest openings; distributed through the mountain ranges of Washington, extending into adjacent British Columbia.

Cascade Huckleberry flowers *Vaccinium deliciosum*

Cascade Huckleberry fruits *Vaccinium deliciosum*

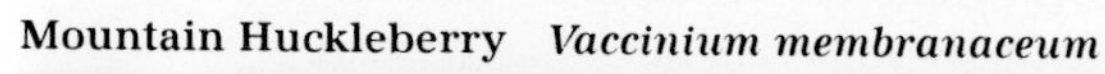

Mountain Huckleberry *Vaccinium membranaceum*

Red Huckleberry

Vaccinium parvifolium

The most common of the lowland blueberries, this is a medium-sized, stiffly erect, deciduous shrub, up to 9 feet (3 m) tall. The stems are green and prominently angled.

Leaves: Alternate, ovate to elliptical, up to 1 inch (2.5 cm) long, sometimes toothed, especially on young plants.

Flowers: Broadly urn shaped with five petal lobes, about ⅛ inch (4 mm) long, white to yellowish pink; sepal lobes well developed.

Fruits: Bright red, tart, juicy berries.

Ecology: Prefers the moist, western side of the mountain ranges, from southeastern Alaska through Washington, extending from lowlands up to high elevations in coniferous forests, typically growing on rotting logs and stumps.

Red huckleberry is easily distinguished from other species of *Vaccinium* by its strongly angled green stems and, especially, by its bright red berries, which are tart but definitely palatable. The flowers are somewhat inconspicuous, hidden by the leaves.

Bog Blueberry

Vaccinium uliginosum

Because of the abundance of bog blueberry in the north, its delicious berries, the persistence of the berries on the branches, and the fact that other food sources are scarce, bog blueberry is a valuable food source for wild animals and humans alike. It is a low, spreading, deciduous shrub, growing about 1 foot (30 cm) high. The stems are yellowish green to reddish and not angled.

Leaves: Alternate, ovate, about 1 inch (2 to 3 cm) long, rather leathery and strongly veined, somewhat grayish.

Flowers: Urn shaped with four petal lobes, longer than wide, pink, typically with broad white stripes; sepal lobes well developed.

Fruits: Juicy and very palatable blue berries.

Ecology: Common in the tundra and black spruce forests (muskegs) of Alaska, especially in bogs; farther south, in British Columbia, it extends upward into the subalpine and alpine zones.

Red Huckleberry *Vaccinium parvifolium*

Bog Blueberry flowers *Vaccinium uliginosum*

Bog Blueberry fruits *Vaccinium uliginosum*

Grouseberry

Vaccinium scoparium

Grouseberry berries are sweet and tasty but are too small and few to excite the serious berry picker. Undoubtedly, they are a select food source for grouse and other animals. A low, matted, deciduous shrub, it grows 1 foot (30 cm) high. The branches are green, broomlike, strongly angled, and strictly erect.

Leaves: Alternate, ovate to lance shaped, small, about ¼ to ½ inch (8 to 16 mm) long, thin, and finely toothed.

Flowers: Urn shaped, with five petal lobes, pinkish, tiny, about ⅛ inch (4 mm long); sepal lobes indistinct.

Fruits: Small, juicy, bright red berries.

Ecology: Grouseberry is probably the most drought-tolerant species of *Vaccinium;* resident on the eastern slopes of the Washington Cascades, ranging northward into southern British Columbia and eastward into the Rockies; common in lowland pine and Douglas fir forests, extending up to and above the treeline; frequently forms a dense ground cover.

Dwarf bilberry *(V. myrtillus)* is a low shrub that closely resembles grouseberry in general form and ecology. It has somewhat larger flowers and leaves, is more hairy, and is less broomlike. The berries are less juicy and darker than those of grouseberry.

Lingon-Berry

Vaccinium vitis-idaea

Lingon-berry is a distinctive dwarf, mat-forming, evergreen shrub.

Leaves: Alternate, leathery, about ¼ to ½ inch (1 to 2 cm) long, typically arched, with a prominent, sunken midvein on top; glandular hairs beneath.

Flowers: Bell shaped with four to five petal lobes, nodding, white to pinkish; sepal lobes prominent and sharp pointed.

Fruits: Bright red, shiny berries; tart but palatable.

Ecology: Widespread in the Arctic tundra and black spruce forests (muskegs), ranging from Alaska southward and westward into British Columbia, where it extends upward through coniferous forests into the alpine zone.

This attractive shrub grows throughout the Arctic region in Eurasia and North America. In northern Scandinavia, the berries have long been prized for preserves and pies.

Grouseberry *Vaccinium scoparium*

Lingon-Berry flowers
Vaccinium vitis-idaea

Lingon-Berry fruits
Vaccinium vitis-idaea

Crowberry Family

Empetraceae

Crowberry is the only species of this family that occurs in the Pacific Northwest.

Crowberry

Empetrum nigrum

Crowberry is a low, spreading to matted evergreen shrub with woolly stems. It closely resembles some heathers but is distinguished by its black berries and a lack of showy flowers.

Leaves: Evergreen, alternate but often appearing whorled because of dense crowding; linear and needlelike, about ¼ inch (4 to 8 mm) long, grooved on the underside.

Flowers: Very small and inconspicuous, petals absent, sepals purplish and enclosed by small bracts that resemble the sepals; stamens three, pistil one.

Fruits: Leathery, dark purple to black berries; unpalatable.

Ecology: Common in bogs and muskegs of southeastern Alaska and northern British Columbia; mainly an alpine cushion plant in Washington and southern British Columbia.

Ginseng Family

Araliaceae

The ginseng family is large, consisting mainly of tropical plants. Only one species, **devil's club,** grows in the Pacific Northwest.

Devil's Club

Oplopanax horridum

The Latin *horridum* is aptly applied to this spiny devil. It produces ginseng-like chemicals, however, that reputedly have healing powers. This shrub has stems that are thick but weak and sprawling branches often more than 6 feet (2 m) long. The plants are densely spiny throughout.

Leaves: Large, up to 1 foot (30 cm) wide, palmately lobed, maplelike; the major veins spiny beneath.

Flowers: Greenish white, small, borne in dense, cone-shaped clusters at branch tips.

Fruits: Bright red berries.

Ecology: A plant of moist to wet forests, especially along streams; ranges from lowlands to middle elevations on the western side of the Cascades and coastal mountains in the Pacific Northwest.

Crowberry, alpine cushion habit *Empetrum nigrum*

Crowberry *Empetrum nigrum*

Devil's Club *Oplopanax horridum*

Honeysuckle Family — Caprifoliaceae

This is a family of shrubs and vines with opposite leaves. The flowers are mainly tubular or funnel shaped, sculpted by the fused petals. The ovary is beneath the other flower parts (inferior) and matures into a berry or capsule.

Twinflower — *Linnaea borealis*

Twinflower is undoubtedly one of the most desirable forest wildflowers, combining beauty with a delightful fragrance. It is a low, creeping, evergreen shrub, often carpeting the forest floor. The flowering stems stand erect, up to 6 inches (15 cm) high.

Leaves: Opposite, leathery and shiny, ovate, usually toothed at the tip, up to 1 inch (2.5 cm) long.

Flowers: Bell shaped and nodding, about ½ to ¾ inch (1 to 2 cm) long, lavender, borne in pairs at the stem tip.

Fruits: Capsules.

Ecology: A common, often dominant understory species in moist conifer forests and bogs, ranging from lowlands to high up in the mountains throughout the Pacific Northwest.

Twinberry — *Lonicera involucrata*

This medium-sized, deciduous shrub stands more or less erect and may grow to be more than 10 feet (3 m) tall.

Leaves: Opposite, elliptical, not toothed, 2 to 6 inches (5 to 15 cm) long and about half as wide.

Flowers: Yellow, funnel shaped, with five petal lobes and a small basal nectar pouch; borne in pairs in leaf axils.

Fruits: Dark purple to black unpalatable berries; two bracts below the berries become large, fleshy, and purplish.

Ecology: Prefers moist to wet soils in open forests, meadows, and lake shores; found throughout the Pacific Northwest, from lowlands to high elevations in the mountains.

Utah honeysuckle *(L. utahensis)*, a related species, is easily distinguished by its red berries and lack of pigmented bracts. It grows in Washington and southern British Columbia.

Twinflower habitat *Linnaea borealis*

Twinflower *Linnaea borealis*

Twinberry *Lonicera involucrata*

Red Elderberry *Sambucus racemosa*

A large, deciduous shrub or small tree, red elderberry has weak, spreading, sprawling branches.

Leaves: Opposite, pinnately compound, the five to seven leaflets sharply toothed and up to 6 inches (15 cm) long.

Flowers: Small, saucer shaped, white, produced in a cone-shaped, branched inflorescence; petals barely fused at the base.

Fruits: Bright red, astringent berries.

Ecology: Grows as a large spreading shrub in moist, open, lowland forests and as a medium-sized shrub in mountain meadows; distributed throughout the Pacific Northwest.

Blue elderberry *(S. cerulea)* is a large shrub or small tree of the eastern Cascades in Washington and the Coast Mountains in southern British Columbia, especially common along major streams and rivers. It resembles red elderberry except it has a flat-topped inflorescence and blue berries.

Highbush Cranberry *Viburnum edule*

A medium-sized, deciduous shrub, highbush cranberry stands more or less erect and grows up to 8 feet (2.5 m) tall.

Leaves: Opposite, sharply toothed, 2 to 4 inches (4 to 10 cm) long, divided at the tip into three pointed lobes.

Flowers: Small, about ⅛ to ¾ inch (4 to 8 mm) across, white, borne in umbrella-like clusters at branch tips; petals fused at the base, with five spreading lobes.

Fruits: Reddish orange berries, tart but edible.

Ecology: A wide-ranging species across northern North America in moist forests and muskegs; distributed throughout the Pacific Northwest but uncommon in Washington and southern British Columbia.

Two species of **snowberry** *(Symphoricarpos)* grow in the Pacific Northwest, mainly in the foothills on both sides of the Washington Cascades and the Coast Mountains of British Columbia. The flowers of both species are small, pale lavender, and funnel shaped or bell shaped. The berries are waxy white and unpalatable (but edible). These shrubs form dense thickets and are important ecologically because of their abundance and because their berries persist into winter, providing survival fare for animals.

Red Elderberry flowers *Sambucus racemosa*

Red Elderberry fruits *Sambucus racemosa*

Highbush Cranberry *Viburnum edule*

Lily Family

Liliaceae

Most members of the lily family are easy to recognize because of their floral characteristics. With few exceptions, the flower parts come in threes–three petals, three sepals, six stamens, and three compartments in the superior ovary (located above the other flower parts). In most species, the petals and sepals are similar in size and color, and the leaves usually have parallel veins. The lily family is large and diverse, with many representatives in the Pacific Northwest.

Wild Onions

Allium species

Wild onions are perennial herbs with underground bulbs. The leaves are linear or lance shaped, occasionally flattened, and basal or nearly so. The flowers have three sepals and three petals, similar in color and shape. They are borne in dense umbels at the stem tip. The fruit in a many-seeded capsule. All species have the characteristic onion taste and odor.

Wild onions usually grow in rocky, exposed sites, mainly at low elevations on both sides of the mountain ranges.

Three onions are common in the region. **Hooker's onion** *(A. acuminatum)* is common on the eastern slopes of the Washington Cascades and occasional on Vancouver Island. It has one to three linear basal leaves and rose-purple flowers. **Nodding onion** *(A. cernuum)* is common in Washington and southern British Columbia, mainly at low elevations. It has several linear basal leaves and nodding, pink flowers. **Scalloped onion** *(A. crenulatum)* is occasional in the Olympics, on the western slopes of the Washington Cascades, and in the Coast Mountains of southern British Columbia. Pink flowers and two flattened leaves spreading from the stem characterize it.

Queen's Cup

Clintonia uniflora

Queen's cup, a low perennial herb, spreads by rhizomes.

Leaves: Basal, two to three, 3 to 8 inches (7 to 20 cm) long, and one-third as wide.

Flowers: Solitary at the tip of leafless stems, white, broadly bell shaped, sepals and petals about ⅝ inch (2 cm) long.

Fruits: Round to oblong berries, blue when mature.

Ecology: A shade-tolerant plant that inhabits moist conifer forests of Washington and southern British Columbia, often as an understory dominant species; ranges upward from lowlands to the treeline.

Hooker's Onion *Allium acuminatum*

Nodding Onion *Allium cernuum*

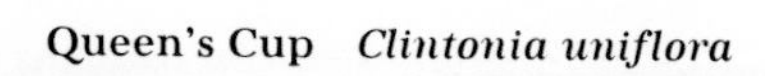

Queen's Cup *Clintonia uniflora*

Hooker's Fairy-Bell *Disporum hookeri*

An adaptable species, this perennial herb has leafy, branched stems 1 to 3 feet (30 to 90 cm) tall.

Leaves: Alternate, ovate to lance shaped, 2 to 6 inches (5 to 15 cm) long, conspicuously parallel veined, upper surface hairy.

Flowers: Creamy white, narrowly bell shaped, usually borne in pairs at branch tips; sepals and petals similar, about ½ inch (12 mm) long, stamens somewhat longer.

Fruits: Orange to red berries, irregularly shaped.

Ecology: Equally at home in moist forests or subalpine meadows; common in the mountains of Washington and southern British Columbia.

Two similar species grow in the Pacific Northwest: **fairy lantern** *(D. smithii)* and **rough-fruited fairy-bell** *(D. trachycarpum)*. The first has larger but narrower (less flaring) flowers and stamens that do not extend beyond the sepals and petals. The second has flowers similar to Hooker's fairy-bell, but the upper surface of the leaves is nonhairy, and it grows on the eastern side of the Cascades in Washington and adjacent British Columbia.

False Lily-of-the-Valley *Maianthemum dilatatum*

An attractive plant, this perennial herb often forms dense populations as a result of rhizomatic growth, often to the exclusion of other plants. The stems are 4 to 12 inches (10 to 30 cm) high.

Leaves: Alternate, two to three per stem, 2 to 4 inches (5 to 10 cm) long, broadly heart shaped, shiny, and conspicuously veined.

Flowers: White, small and numerous in a narrow raceme or spike at the stem tip; flower parts come in fours, which is unusual in the lily family; four sepals or petals and four stamens.

Fruits: Red berries.

Ecology: A locally abundant species in moist to wet forests and on stream banks, from lowlands to mid-elevations on the western side of the Cascades and Olympics in Washington, extending northward into Alaska along the coastal ranges.

Hooker's Fairy-Bell *Disporum hookeri*

Rough-Fruited Fairy-Bell *Disporum trachycarpum*

False Lily-of-the-Valley *Maianthemum dilatatum*

Glacier Lily *Erythronium grandiflorum*

Glacier lily is a favorite among wildflower enthusiasts. It grows in abundance in subalpine meadows, where its beautiful, yellow flowers appear earlier in spring than flowers of most associated plants. As snowfields recede up the mountains, it blooms progressively later. Also called dogtooth violet, this perennial herb grows from an elongate bulb, with stems up to 1 foot (30 cm) high.

Leaves: Two to three per plant, basal or nearly so, elliptical, 4 to 8 inches (10 to 20 cm) long, one-fourth as wide, rather leathery and shiny.

Flowers: Pale to dark yellow, showy, one to three per plant, nodding; sepals/petals about 1 inch (2 to 3 cm) long and typically reflexed backward, exposing the stamens and style.

Fruits: Erect, elongate capsules.

Ecology: A variable species that can grow in open forests, subalpine meadows, and even the sagebrush steppe; ranges from southern British Columbia through the mountains of Washington.

Fawn Lily *Erythronium montanum*

Also called avalanche lily, this perennial herb grows from an elongate bulb. The stems are 6 to 10 inches (15 to 25 cm) high.

Leaves: Basal or nearly so, two to three per stem, 4 to 8 inches (10 to 20 cm) long, with a distinct petiole, rather leathery and shiny.

Flowers: Creamy white, becoming pinkish with age, showy, one to three per stem, nodding; sepals/petals 1 to 2 inches (2 to 5 cm) long, spreading or reflexed backward; marked at the base by a yellow band.

Fruits: Erect, elongate capsules.

Ecology: Common in the South Cascades of Washington, occurring sporadically in the North Cascades, Olympics, and mountains of Vancouver Island; prefers moist, open, high-elevation forests and subalpine meadows.

Another attractive lowland species of western Washington and coastal British Columbia is **trout lily** *(E. oregonum).* This charming white-flowered plant has eye-catching, mottled leaves (pale green on dark green).

Glacier Lily
Erythronium grandiflorum

Fawn Lily
Erythronium montanum

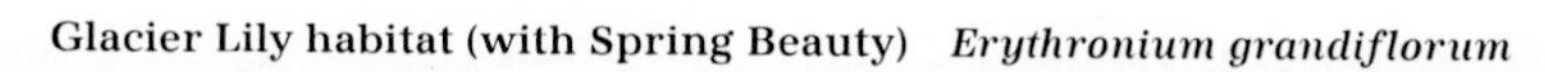

Glacier Lily habitat (with Spring Beauty) *Erythronium grandiflorum*

Tiger Lily *Lilium columbianum*

Tiger lily is a spectacular wildflower. The flowers are large and charmingly shaped, with unusual marking and color. A common perennial herb that grows from a bulb, it has stout stems and grows up to 4 feet (1.2 m) tall.

Leaves: Whorled or alternate, elliptical to lance shaped, the larger ones 2 to 4 inches (5 to 10 cm) long, leathery.

Flowers: Very showy, yellow-orange to reddish with purplish spots, nodding, few to several per plant, borne individually on long stalks from bracts near the stem tip; sepals and petals reflexed backward.

Fruits: Many-seeded capsules.

Ecology: Has a broad ecological tolerance, growing in forests or open areas from lowlands to subalpine meadows; ranges from southern British Columbia through the mountains of Washington.

The *Fritillaria* genus is closely related to *Lilium.* **Checker lily** *(F. affinis),* also known as chocolate lily, is commonly found throughout the region on rocky slopes at low elevations; it occasionally grows in subalpine meadows. Like the tiger lily, it has whorled leaves and a few to several nodding flowers, but the sepals and petals are purplish and mottled with green or yellow, sometimes both. This species is largely replaced in the northern part of the region by black lily *(F. camschatcensis),* which has striped rather than mottled flowers.

Star-Flowered Solomon's Seal *Smilacina stellata*

This small perennial herb spreads by rhizomes. The stems are usually single, unbranched, and stand 1 to 2 feet (30 cm to 0.6 m) high.

Leaves: Alternate, lance shaped, up to 6 inches (15 cm) long, heavily veined.

Flowers: Small and white, a few to several, borne in a zigzag raceme at the stem tip; the sepals and petals spread outward like a star.

Fruits: Greenish to black berries.

Ecology: Usually found in rocky but moist soil along streams, in open forests, and in subalpine meadows; distributed throughout the Pacific Northwest.

A similar but more robust plant is **false Solomon's seal** *(S. racemosa).* It has clustered stems, leaves much larger and wider than star-flowered Solomon's seal, and a branched inflorescence bearing numerous small, white, delightfully fragrant flowers. The two species have similar distributions.

Tiger Lily *Lilium columbianum*

Checker Lily *Fritillaria affinis*

Star-Flowered Solomon's Seal *Smilacina stellata*

False Solomon's Seal *Smilacina racemosa*

Western Mountain Bells *Stenanthium occidentale*

A moisture-loving perennial herb, western mountain bells grows from a small bulb. The stems are unbranched and stand 4 to 14 inches (10 to 36 cm) high.

Leaves: Basal, long and linear, grasslike.

Flowers: Bell shaped, yellowish green to reddish purple, few to several in narrow racemes.

Fruits: Large capsules.

Ecology: Locally common on cliffsides and in rocky seepages and moist meadows in the subalpine zone, occasionally at lower elevations; found in Washington and southern British Columbia.

Bog Lily *Tofieldia glutinosa*

Bog lily is an herbaceous perennial with sticky, clustered stems 4 to 20 inches (10 to 50 cm) tall.

Leaves: Alternate, situated on the lower half of the stem, linear and grass-like, up to 6 inches (15 cm) long.

Flowers: Small, white to greenish, borne in a headlike cluster at the stem tip.

Fruits: Many-seeded capsules.

Ecology: An occasional plant in wet areas throughout the Pacific Northwest, especially bogs, from lowlands to the alpine zone.

Trillium *Trillium ovatum*

The beautiful and fragrant trillium is a perennial herb with fleshy, tuberous roots. The stems are sometimes clustered and may grow up to 1 foot (30 cm) high.

Leaves: Usually three; situated near the stem tip, broadly ovate, 2 to 6 inches (5 to 15 cm) long and wide, net veined.

Flowers: Very showy, solitary at the stem tip; petals three, white, becoming reddish with age, 1 to 2 inches (2.5 to 5 cm) long; much larger than the green sepals.

Fruits: Leathery capsules.

Ecology: Resident in moist forests and along stream banks from lowlands up to the subalpine zone; common in the Coast Mountains of southern British Columbia and in the Washington Cascades and Olympics.

Western Mountain Bells
Stenanthium occidentale

Bog Lily
Tofieldia glutinosa

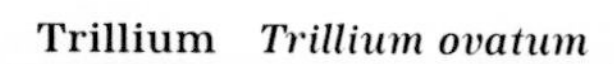

Trillium *Trillium ovatum*

Rosy Twisted-Stalk *Streptopus roseus*

An herbaceous perennial partial to wet habitat, this species has unbranched, leafy stems and stands 6 to 16 inches (15 to 40 cm) high.

Leaves: Alternate, ovate to elliptical, 2 to 4 inches (5 to 10 cm) long and half as wide, pointed and strongly veined.

Flowers: Bell shaped, rose to purplish, ¼ to ⅜ inch (6 to 10 mm) long, borne from leaf axils, one flower per leaf. The flowers are borne on twisted stalks that position them under the leaves, hiding them from view from above.

Fruits: Red berries.

Ecology: A common plant in moist conifer forests, in forest openings, and along streams from lowlands to the subalpine zone in southeastern Alaska and southward through Washington.

Clasping-leaved twisted-stalk *(S. amplexifolius)* grows in the same region. It is less common than rosy twisted-stalk, especially at higher elevations, but is more widely distributed, growing throughout the Pacific Northwest. It differs from rosy twisted-stalk by having branched stems and leaves that clasp the stem. Also, the flowers are white to greenish, and the sepals and petals flare at the tips.

Beargrass *Xerophyllum tenax*

This plant is also called Indian basket grass because the tough, fibrous leaves were used by Native Americans to weave baskets. The name "beargrass" probably relates to the unpleasant bearlike odor of the flowers. A particularly stout, herbaceous perennial, beargrass has stems 2 to 5 feet (0.6 to 1.5 m) tall, covered by wiry leaves.

Leaves: Basal leaves densely tufted, linear, wiry, and up to 2 feet (0.6 m) long; stem leaves progressively shorter upward but equally wiry.

Flowers: White, individually small, but borne in a dense, showy conelike cluster at the stem tip.

Fruits: Small capsules.

Ecology: A conspicuous plant, often an understory dominant species, in open, dry forests on the eastern side of the Washington Cascades and Olympics.

Rosy Twisted-Stalk *Streptopus roseus*

Clasping-Leaved Twisted-Stalk *Streptopus amplexifolius*

Beargrass *Xerophyllum tenax*

Green False Hellebore

Veratrum viride

A robust perennial herb that spreads by rhizomes, with stems up to 6 feet (2 m) tall.

Leaves: Alternate, broadly ovate like a cabbage leaf, 6 to 12 inches (15 to 30 cm) long and nearly as wide, strongly veined.

Flowers: Yellowish green, numerous on narrow, typically drooping branches near the stem tip; sepals and petals about ¼ inch (5 to 10 mm) long.

Fruits: Capsules.

Ecology: Found in wet meadows and marshes throughout the Pacific Northwest from lowlands to the subalpine zone, where it grows on slopes well watered from melting snow.

A related species with a more southerly distribution but common in Washington is **California false hellebore** *(V. californicum).* This plant has more erect branches of whitish rather than yellowish green flowers. Like green false hellebore, it is highly poisonous.

Mountain Death-Camas

Zygadenus elegans

Mountain death-camas, another poisonous perennial herb, grows from a bulb. The stems are unbranched and ½ to 2 feet (15 to 60 cm) high.

Leaves: Basal or nearly so, linear and grasslike, up to 1 foot (30 cm) long.

Flowers: Greenish white, in a narrow raceme at the stem tip; sepals and petals about ⅜ inch (8 to 10 mm) long, bearing a green spot (gland) at the base.

Fruits: Many-seeded capsules.

Ecology: Grows in open forests, meadows, and moist rocky slopes from lowlands into the subalpine zone; common in Alaska and northern British Columbia, becoming rare in Washington.

In Washington and southern British Columbia, **meadow death-camas** *(Z. venenosus)* is much more common, but its distribution is restricted to the foothills on both sides of the mountain ranges. Mountain death-camas has larger and more attractive flowers (*elegans* means "elegant"). Both species are highly poisonous.

Green False Hellebore *Veratrum viride*

Green False Hellebore
Veratrum viride

Mountain Death-Camas
Zygadenus elegans

MADDER FAMILY — Rubiaceae

Rubiaceae is a large and variable family, primarily tropical but with a few representatives in western North America, mainly in the genus *Galium*. The genus is typified by northern bedstraw.

Northern Bedstraw — *Galium boreale*

Northern bedstraw (*boreale* means "northern") is a perennial herb that spreads by creeping rhizomes. The stems are erect and 8 to 20 inches (20 to 50 cm) tall.

Leaves: Borne in whorls of four, lancelike to linear, about 1 inch (2 to 4 cm) long, three veined.

Flowers: Small, white, parts in fours (four sepals, four petals, four stamens), borne in stalked clusters from upper leaf axils; petals are fused and funnel shaped, with four spreading lobes.

Fruits: Small, paired, one-seeded nutlets, lacking hooked bristles.

Ecology: Widespread and variable, distributed throughout the Pacific Northwest; occurs in a variety of habitats, from open forests to meadows, ranging from lowlands upward into the alpine zone.

MOUNTAIN BOX (STAFF-TREE) FAMILY — Celastraceae

The mountain box family is large and variable, but only one species grows in the Pacific Northwest.

Mountain Box — *Pachistima myrsinites*

In Douglas fir and pine forests, this attractive shrub is frequently an understory dominant, often in association with grouseberry, which it superficially resembles. Also known as Oregon boxwood, it is a low, extensively branched, evergreen shrub, seldom more than 2 feet (0.6 m) high.

Leaves: Opposite, leathery, toothed, ovate, 3/8 to 1¼ inches (1 to 3 cm) long.

Flowers: Small and inconspicuous, about 1/8 inch (4 mm) broad, borne in leaf axils; petals four, maroon.

Fruits: Small capsules, about 1/8 inch (3 to 4 mm) long.

Ecology: Prefers dry, open forests, ranging from middle to high elevations, mainly on the eastern side of the Cascades in Washington and adjacent British Columbia.

Northern Bedstraw *Galium boreale*

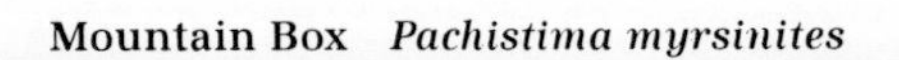

Mountain Box *Pachistima myrsinites*

Mustard Family
Brassicaceae (Cruciferae)

Plants in the mustard family typically have flowers with four petals that form the shape of a cross, from which the alternate family name, Cruciferae, is derived. Apart from the flowers, and the mustard oils that produce a radishy taste sensation, plant species in this family vary considerably, from diminutive annuals to shrubs. Many members of the family are weeds, and some species are highly invasive. Besides the characteristic four petals, the flowers also have four sepals and usually six stamens, two short and four long. The fruit is a two-compartmented pod, the halves separating at maturity, sometimes explosively, releasing the seeds and leaving a central, papery partition.

Lemmon's Rockcress
Arabis lemmonii

One of several alpine rockcresses, this low perennial herb usually has several spreading stems 2 to 6 inches (5 to 15 cm) long, derived from a branched, woody root crown. Minute, branched hairs cover the plant.

Leaves: Basal leaves several, elongate, wider near the tip and usually toothed; stem leaves ovate to lance shaped.

Flowers: Showy, few to several in short racemes at the stem tips; petals four, rose-purple, about ¼ inch (5 to 7 mm) long.

Fruits: Narrow, linear spreading pods, 1 to 2 inches (2 to 5 cm) long.

Ecology: An alpine species, it grows on rocky ridges, along cliffsides, and on talus slopes in the mountains of Washington and southern British Columbia.

Two similar rockcress species, *A. lyallii* and *A. microphylla,* grow in the Pacific Northwest. All are low, more or less cushion plants, with rose-purple flowers. Their habitats are similar and they overlap in distribution.

Payson's Draba
Draba paysonii

A very densely matted, hairy herb, Payson's draba grows only ⅜ to 1¼ inches (1 to 3 cm) high.

Leaves: Elongate, about ¼ to ⅜ inch (5 to 10 mm) long, crowded on the lower stem.

Flowers: Bright yellow, few to several on each of the numerous short stems; petals four, about ⅛ inch (2 to 5 mm) long.

Fruits: Small, ovate, flattened, few-seeded pods.

Ecology: Exposed rocky sites in the subalpine and, especially, alpine zones; common in Washington and British Columbia.

Other yellow-flowered, cushion drabas grow in the region, in similar habitats. The most common of these attractive plants is **few-seeded draba** *(D. oligosperma).* Differentiating among the various species is difficult.

Lemmon's Rockcress *Arabis lemmonii*

Payson's Draba *Draba paysonii*

Few-Seeded Draba *Draba oligosperma*

Lance-Fruited Draba *Draba lonchocarpa*

A low, densely hairy, perennial herb, this draba has clustered, mainly leafless stems up to 6 inches (15 cm) high.

Leaves: Basal or sometimes one at the base of the stem, linear to ovate, ¼ to ⅝ inch (5 to 15 mm) long.

Flowers: Small, few to several in short racemes at the stem tip; petals white, about ⅛ inch (3 to 5 mm) long.

Fruits: Narrow, elongate pods, about ¼ to ½ inch (5 to 15 mm) long.

Ecology: A common though inconspicuous plant in rocky alpine sites, this highly variable species grows throughout the Pacific Northwest in one form or another.

Mountain Wallflower *Erysimum arenicola*

Mountain wallflower is a perennial herb with one to several erect stems up to 1 foot (30 cm) high. The plants are hairy overall.

Leaves: Basal and alternate, about 1 inch (2.5 cm) long including the petiole, longer than wide and widest near the tip, usually toothed.

Flowers: Showy, lemon yellow, densely clustered in a terminal raceme; petals about 1 inch (2 cm) long.

Fruits: Linear pods, 1 to 4 inches (3 to 10 cm) long, with a beaklike tip, enlarged where the seeds occur.

Ecology: Rocky ridgetops, talus slopes, and rock crevices in the high mountains of Washington and Vancouver Island.

The *torulosum* variety, which often has single stems, best represents this variable species in the Pacific Northwest.

Smelowskia *Smelowskia calycina* and *S. ovalis*

These two high-elevation mustards are densely matted perennial herbs with several stems 2 to 8 inches (5 to 20 cm) long. Soft, gray hair covers the plants. The species are named for a nineteenth-century Russian botanist, Timotheus Smelowski.

Leaves: Basal and alternate; pinnately compound; ¾ to 2½ inches (2 to 6 cm) long.

Flowers: Small, white, clustered in short, terminal racemes.

Fruits: Elliptical, flattened pods, ⅛ to ¼ inch (3 to 6 mm) long.

Ecology: The smelowskias are alpine plants of rock crevices and talus slopes in the mountains of Washington and southern British Columbia.

Lance-Fruited Draba *Draba lonchocarpa*

Mountain Wallflower *Erysimum arenicola*

Smelowskia *Smelowskia calycina*

Orchid Family

Orchidaceae

By most estimates, the orchids comprise the second largest family of flowering plants, after the sunflower family. It is also one of the most interesting plant families, for several reasons. In most cases, the species are specialists in terms of pollination strategy, habitat, and association with fungi. The flowers are bilaterally symmetrical, highly ornate, and recognizably unique, characteristics leading to pollination specialization. Because the species are habitat specialists, they must have some way of finding their specific habitat type. The production of thousands or even millions of seeds per flower ensures their success. The seeds are minute, and capable of being blown great distances in the wind. To fertilize so many seeds, huge numbers of pollen grains must be deposited on the stigmas. This is accomplished by having millions of pollen grains transported as a unit by a single insect. In exchange for high mobility, the plant must sacrifice food reserves that would enhance seed security. This is where the association with fungi plays a role in the plant's success. A specific fungus "attacks" an orchid seed, but the seed feeds on the fungus, deriving the energy necessary for seedling establishment. Some nongreen orchids never outgrow their dependency on their associated fungus.

Most orchids are tropical, growing in the forest canopy. In the Pacific Northwest, the family is not well represented, and the flowers usually are not particularly showy. However, all orchid species have flowers that are ornately sculpted through modification of the three sepals and, especially, the three petals. The stamens are fused to the style, and the ovary is inferior, or located below the other flower parts. All orchids are perennials.

Fairy Slipper

Calypso bulbosa

Fairy slipper is a small orchid with a single, leafless stem that grows from a small bulb. Unfortunately, the shallow bulb is regularly pulled up when the attractive flower is thoughtlessly picked. Thus a life requiring years to develop is wasted.

Leaves: Solitary, derived from the bulb, the blade 1¼ to 2½ inches (3 to 6 cm) long.

Flowers: Very showy, single at the stem tip; the three sepals and two petals are alike–ascending and lavender with darker stripes; the other petal is enlarged, forming the pendant, saclike lip, which is white with reddish purple and yellow markings.

Fruits: Capsules with thousands of minute seeds.

Ecology: An occasional to locally abundant plant in moist, cool forests, usually on humus-rich soils from low to mid-elevations throughout the region.

A larger **lady-slipper** *(Cypripedium montanum)* also occurs in the region, on the eastern side of the Cascades in Washington and adjacent British Columbia. It has brownish purple sepals and matched petals, and a white, much enlarged, saclike lip. It has leafy stems, the leaves up to 6 inches (15 cm) long and conspicuously veined.

Fairy Slipper *Calypso bulbosa*

Lady-Slipper *Cypripedium montanum*

Spotted Coral Root *Corallorhiza maculata*

This orchid, a succulent herb, has purplish to reddish brown stems that are either single or clustered, 6 to 18 inches (15 to 45 cm) tall.

Leaves: Reduced to sheathing bracts the color of the stems.

Flowers: Individually showy, about ½ inch (1 to 2 cm) long, including the inferior ovary, several borne in a narrow raceme at the stem tip; sepals reddish purple or yellowish; petals shorter than the sepals—the two uppermost are pinkish with darker stripes, and the lower (the lip) is white with maroon spots.

Fruits: Capsules with thousands of minute seeds.

Ecology: A rather common plant in conifer forests from lowlands to mid-elevations throughout the mountains in Washington and southern British Columbia.

Albino forms of spotted coral root are common, often interspersed with the other plants. They are uniformly pale yellow, including the flower, except for the white, nonspotted lip.

Western Coral Root *Corallorhiza mertensiana*

This is the most common coral root orchid. It has reddish, solitary or clustered stems 6 to 18 inches (15 to 45 cm) tall.

Leaves: Reduced to sheathing bracts the same color as the stem.

Flowers: Small, few to several in a narrow raceme at the stem tip; sepals narrow and pinkish; the two upper petals similar to sepals, the lower petal (the lip) pink, typically with white splotches and dark pink veins. A narrow spur projects outward from below the lip.

Fruits: Capsules with thousands of minute seeds.

Ecology: A plant of middle to high elevations in humus-rich, conifer forests in Washington and southern British Columbia.

Like spotted coral root, this species has albino forms. They are smaller and have a more prominent flower spur.

Another species resembling western coral root is **striped coral root** *(C. striata)*. It has larger, showier flowers with purplish red stripes on the otherwise white to yellowish or pale pink sepals and petals, including the lip. Also, the sepals and petals converge toward their tips.

Spotted Coral Root *Corallorhiza maculata*

Western Coral Root *Corallorhiza mertensiana*

Western Coral Root *Corallorhiza mertensiana*

Striped Coral Root *Corallorhiza striata*

Rattlesnake Plantain *Goodyera oblongifolia*

Rattlesnake plantain is a perennial herb with evergreen leaves. It spreads by rhizomes. The stems are solitary, about 1 foot (30 cm) high, and are sticky.

Leaves: All basal (plantainlike), conspicuously mottled or striped with white (like a rattlesnake); otherwise dark green, thick, and leathery; blades elliptical, 1 to 3 inches (3 to 7 cm) long.

Flowers: Small, greenish white, several in a narrow raceme along the upper half of the stem, tending to face in one direction; upper sepal and two petals project forward, forming a hood over the lip; the lateral two sepals are rolled outward.

Fruits: Capsules with thousands of minute seeds.

Ecology: A common resident in conifer forests from low to high elevations in the mountains in Washington and British Columbia, barely reaching into southeastern Alaska.

Heart-Leaved Twayblade *Listera cordata*

This is a small, delicate plant with slender, creeping rhizomes. The stems are solitary, 3 to 8 inches (7 to 20 cm) high, and are sticky near the top.

Leaves: Borne in pairs ("tway") near the middle of the stem, usually heart shaped, ½ to 1½ inches (1 to 4 cm) long.

Flowers: Small, pale green to purplish, few to several in a terminal raceme; the lower petal (the lip) is split halfway, the narrow lobes diverging.

Fruits: Capsules with thousands of minute seeds.

Ecology: A common though inconspicuous plant in moist conifer forests, preferring a decaying wood substrate; distributed throughout the Pacific Northwest, from lowlands to high elevations in the mountains.

Other species of *Listera* grow in similar habitats in the region, differing from heart-leaved twayblade in that the flower lip, though often notched, is not as deeply dissected. The most common of these is **northwestern twayblade** *(L. caurina),* which has greenish flowers with a rounded or slightly notched lip. The flowers fancifully resemble a cow's head with horns and erect ears (the sepals and upper petals) and a projected snout (the lip). The leaves are ovate, rather thick, and 1 to 2 inches (2.5 to 6 cm) long.

attlesnake Plantain *Goodyera oblongifolia*

Rattlesnake Plantain *Goodyera oblongifolia*

Heart-Leaved Twayblade *Listera cordata*

Northwestern Twayblade *Listera caurina*

White Bog Orchid *Platanthera (itabenaria) dilatata*

White bog orchid is one of the most admired wildflowers, with ornate and delightfully fragrant waxy white flowers. It often forms dense communities in mountain bogs. A rather succulent perennial herb, it grows from thick, fibrous roots. The stems are leafy, hairless, and up to 3 feet (1 m) tall.

Leaves: Alternate, thick and leathery, lance shaped, 2 to 4 inches (4 to 10 cm) long, their base sheathing the stem.

Flowers: White, individually small but showy in a dense terminal cluster; the upper sepal and two petals converge at the tip to form a hood; the lower two sepals are spreading; the lip is broad at the base and narrow at the tip; behind the lip, and of nearly equal length, hangs a narrow nectar tube called a "spur."

Fruits: Capsules with thousands of minute seeds.

Ecology: As the common name suggests, this species grows in wet places–bogs, seepages, stream banks, and wet meadows. Distributed throughout the region, ranging from lowlands upward into the subalpine zone.

Several other species of *Platanthera* grow in the Pacific Northwest, and most of them are equally partial to wet, boggy soils. Although they are often difficult to tell apart, they are distinct from white bog orchid in having greenish, less showy flowers. A common representative of this green-flowered group is **slender bog orchid** *(P. stricta).* It has small flowers in a narrow, open spike. The spur, short and enlarged at the tip, characterizes the species.

Ladies' tresses *(Spiranthes romanzoffiana)* superficially resembles white bog orchid, with thick, lance-shaped leaves and a dense terminal cluster of small, white flowers. The plants are generally smaller, however, and the leaves are restricted to the lower half of the stem; also, the flowers lack spurs. The most obvious distinction relates to the inflorescence–the flowers are spirally arranged in an attractive spike, an arrangement responsible for both the common and generic names. Also, like white bog orchid, this species grows in wet, swampy, or boggy areas throughout the Pacific Northwest, mainly in mountain meadows and forest openings.

White Bog Orchid *Platanthera dilatata*

Slender Bog Orchid *Platanthera stricta*

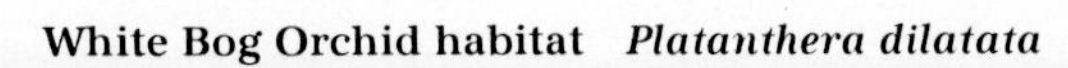

White Bog Orchid habitat *Platanthera dilatata*

Oleaster Family

Elaeagnaceae

A small family of woody plants, with only one Pacific Northwest species, buffalo berry.

Buffalo Berry

Shepherdia canadensis

This medium-sized, deciduous, spreading shrub grows 3 to 10 feet (1 to 3 m) tall. Brownish branlike "scabs" mark the stems and leaves.

Leaves: Opposite, ovate to elliptical, 1 to 2 inches (2 to 6 cm) long, dark green with white spots on the upper surface, the undersurface covered with white hairs and rusty brown, branlike scales.

Flowers: Unisexual, borne at the base of leaves, inconspicuous with no petals; sepals four, stamens eight (in male flowers).

Fruits: Round to oblong berries, reddish, edible but bitter.

Ecology: Common in southeastern Alaska and northern British Columbia and uncommon farther south. Resident in open forests and thickets from lowlands to mid-elevations.

Parsley Family

Apiaceae (Umbelliferae)

The parsley family is large, but in the Pacific Northwest most native species are restricted to dry habitats on the eastern side of the mountains. The two outstanding characteristics of the family are parsleylike leaves and a flower cluster that is a compound umbel (umbels in umbels)–the stalks of the secondary, terminal umbels resemble ribs of an umbrella. The individual flowers are small, with five petals, usually no sepals, five stamens, and a two-seeded, inferior ovary that splits into two, one-seeded units at maturity.

Cow Parsnip

Heracleum lanatum

A conspicuous and odiferous plant, cow parsnip is a robust, woolly, perennial herb. The stems are solitary, thick, and hollow, standing 3 to 10 feet (1 to 3 m) tall.

Leaves: Very large, divided into three palmately lobed leaflets, which are as much as 1 foot (30 cm) wide; petioles are enlarged and sheath the stem.

Flowers: Small and white, numerous, in one or more large, flattened, compound umbels; the outer petals of the umbels are enlarged.

Fruits: Ovate and flattened; splits at maturity into two, one-seeded "nutlets."

Ecology: Inhabits stream banks and other moist to wet places, including subalpine meadows; occurs throughout the Pacific Northwest.

Buffalo Berry *Shepherdia canadensis*

Cow Parsnip *Heracleum lanatum*

Cow Parsnip *Heracleum lanatum*

Gray's Lovage *Ligusticum grayi*

A stout taproot supports this perennial, hairless herb. The stems are solitary or few, up to 2 feet (0.6 m) tall.

Leaves: Several times divided and parsleylike, basal leaves well developed with long petioles; stem leaves are smaller and lack petioles.

Flowers: Small, white to purplish, numerous in flattened or round-topped, compound umbels, secondary umbels in a primary umbel.

Fruits: Elliptical with prominent ribs; splits at maturity into two, one-seeded "nutlets."

Ecology: Locally abundant on open to wooded slopes and in subalpine meadows of the Cascades in Washington and adjacent British Columbia.

Few-Flowered Desert Parsley *Lomatium martindalei*

A low, often prostrate herb, it grows from an elongate taproot. The plants usually have several spreading to erect stems, up to a few inches (4 to 12 cm) long.

Leaves: Divided and parsleylike, primarily basal, stem leaves smaller.

Flowers: Small, yellow to nearly white, numerous, in stalked, compound umbels; secondary umbels small, roundish, and short stalked.

Fruits: Elliptical, flattened, splits at maturity into two, one-seeded "nutlets."

Ecology: Grows on rock outcrops and rocky ridges in the subalpine and alpine zones; common in the mountains of Washington and southern British Columbia.

This variable species has two varieties in the Pacific Northwest, distinguished largely on the basis of flower color.

Other *Lomatium* species frequently extend upward into the mountains, especially on the dry, eastern slopes. Perhaps the most common of these is **fern-leaf desert parsley** *(L. dissectum),* a robust plant, usually with several flowering stems and large leaves divided into small segments. This species is subdivided into several varieties. The Cascade Mountain form (var. *dissectum*) ranges southward through Washington and northward into southern British Columbia, growing on dry, south-facing slopes and extending upward into the subalpine zone. It has bright yellow flowers.

Gray's Lovage *Ligusticum grayi*

Few-Flowered Desert Parsley *Lomatium martindalei*

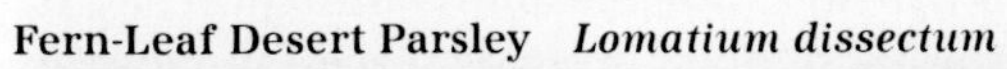

Fern-Leaf Desert Parsley *Lomatium dissectum*

Pea Family Fabaceae (Leguminosae)

Members of the pea family are well known for their ability to fix nitrogen, enriching the soil where they live and increasing the nitrogen and protein content of the plant tissue. This unique feature gives the plants an important dual ecological role–habitat improvement and providing forage for herbivores. Unfortunately, many of the species are deadly poisonous, especially the locoweeds. Legumes also provide a valuable source of pollen and/or nectar for specialized pollinators, particularly bumblebees. Most pea species have beautiful flowers, which do not go unnoticed or unappreciated.

Although there is much variation among the species, uniformity reigns in their sweetpea-like flower structure. The uppermost of five petals is enlarged and is called the **banner;** the two lateral petals are called **wings;** and the two lower petals are partially fused to form the **keel,** so named because of its resemblance to a boat's keel. The ten stamens and the style are entrapped and "spring-loaded" inside the keel. When the keel is depressed by a visiting insect, the stamens and style are released, striking the insect on the abdomen. Pollen is deposited by the anthers and picked up by the stigma–a rather sophisticated pollination strategy. The fruit type is another unifying characteristic among the species. It is a modified pod (a legume) that splits open along two sutures with an energetic twisting motion that pops off and throws the seeds.

The best representative of the pea family in the Pacific Northwest, at least in terms of color display, is the lupine genus, *Lupinus.* Lupines are conspicuous and often dominant members of subalpine meadows throughout the region. Their palmately compound leaves are also distinctive. However, the categorizing of lupines is notoriously difficult. The *Lupinus arcticus* complex, for example, is often expanded to include **broadleaf lupine** *(L. latifolius)* as a subspecies.

Another species that is frequently combined with *L. arcticus* is **large-leaved lupine** *(L. polyphyllus).* As its name implies, large-leaved lupine has larger and more numerous leaflets, usually ten or more, and is more robust. It is common in southern British Columbia and Washington and has a broad ecological tolerance, growing in open to shady places, in moist to dry habitats, and from lowlands to the alpine zone.

Lupines are pollinated by bumblebees that manipulate the flowers to collect pollen, which they package on their hind legs and carry back to the hive to feed larvae. Lupine is one of the wildflowers most favored by outdoor enthusiasts; dense lupine populations form spectacular displays. But herbivores beware–lupines are poisonous.

Broadleaf Lupine *Lupinus latifolius*

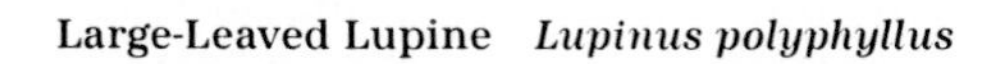

Large-Leaved Lupine *Lupinus polyphyllus*

Arctic Lupine *Lupinus arcticus* ssp. *arcticus*

A northern perennial herb, arctic lupine grows up to 2 feet (0.6 m) tall, with a few to several stems arising from a woody, branched root crown.

Leaves: Mainly basal with long petioles, palmately compound with six to nine sharp-pointed leaflets; leaflets hairy underneath and hairless above.

Flowers: Blue with a banner that is whitish before pollination and red after; many flowers grow in a narrow raceme.

Fruits: Hairy pods.

Ecology: Common throughout much of Alaska, ranging southward through the mountains into British Columbia, where it grows in mountain meadows and high, open forests.

Broadleaf Lupine *Lupinus latifolius*

This is the most common lupine in the region. Broadleaf lupine is similar to arctic lupine but is generally more robust, up to 3 feet (1 m) tall.

Leaves: Similar to those of arctic lupine but the leaflets are usually hairy on both surfaces.

Flowers: Blue to lavender, showy, numerous, in a narrow, spirelike raceme; banner white, becoming red after pollination.

Fruits: Woolly pods.

Ecology: A mountain wildflower of Washington and southern British Columbia, where it inhabits moist subalpine meadows and open forests at high elevations.

Alpine Lupine *Lupinus lyallii*

A prostrate, mat-forming, perennial herb that grows from a branched, woody rootstalk.

Leaves: Mainly basal, palmately compound with five to nine narrowly elliptical, densely hairy leaflets; petioles at least twice as long as the leaflets.

Flowers: Blue to rarely white, clustered in dense but short (headlike) racemes at the stem tips.

Fruits: Short, woolly pods.

Ecology: Grows on rocky, often volcanic slopes in the alpine and subalpine zones in the mountains of Washington and southern British Columbia.

Arctic Lupine *Lupinus arcticus*

Broadleaf Lupine *Lupinus latifolius*

Alpine Lupine *Lupinus lyallii*

Alpine Locoweed

Astragalus alpinus

This is a matted, hairy, perennial herb that spreads by rhizomes. The stems are leafy, 3 to 10 inches (7 to 25 cm) long, and are erect only at the tips.

Leaves: Alternate, pinnately compound with eight to ten pairs of elliptical leaflets, each up to ¾ inch (2 cm) long.

Flowers: Bilaterally symmetrical, sweetpea-like, two-toned white and lavender-purple, crowded in a narrow, terminal raceme.

Fruits: Pods, covered with black hair.

Ecology: Prefers rocky sites in open forests and along ridges from lowlands (in the north) into the alpine zone; common in the northern part of the Pacific Northwest, extending southward into the mountains of central Washington.

Astragalus is the largest genus of flowering plants in western North America. Of the many species, most are called "milkvetches" and "locoweeds" and are distributed over the drier steppes and prairies. Some of the dry-land species extend well up into the mountains on open, south-facing slopes.

Field Crazyweed

Oxytropis campestris

As the common name suggests, this hairy, perennial herb is poisonous. Several stems, 4 to 8 inches (10 to 20 cm) high, grow from a branched rootstalk.

Leaves: Basal, pinnately compound with numerous, narrowly elliptical leaflets.

Flowers: Yellow to cream colored, crowded in a headlike cluster at the tip of erect, unbranched, leafless stems.

Fruits: Hairy pods.

Ecology: Prefers rocky sites in the mountains, especially in the alpine zone; distributed throughout the Pacific Northwest.

Oxytropis campestris is extremely variable, and nomenclatural revisions have been proposed. A less hairy and lower form that grows in southeastern Alaska and northern British Columbia has been given the name *O. varians* (*O. campestris* var. *varians*). The form that grows in Washington and southern British Columbia is *O. campestris* var. *gracilis*. Other species of *Oxytropis* inhabit the region, especially in the drier mountains in southeastern Washington. Some of them have purplish flowers.

Species of *Oxytropis* are similar to those of *Astragalus* but have leafless stems. Species of both genera are poisonous.

Alpine Locoweed *Astragalus alpinus*

Field Crazyweed *Oxytropis campestris* var. *gracilis*

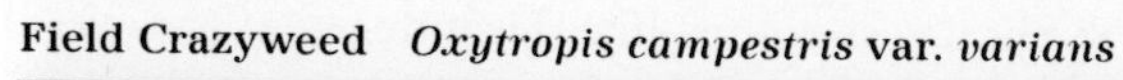

Field Crazyweed *Oxytropis campestris* var. *varians*

Alpine Sweet-Vetch *Hedysarum alpinum*

A perennial herb with a few to several spreading, branched stems, 4 to 18 inches (10 to 45 cm) long, occasionally longer.

Leaves: Alternate, pinnately compound with nine to nineteen leaflets that are more or less conspicuously veined and usually have pointed tips.

Flowers: Dark to pale lavender, sweetpea shaped, about 5/8 inch (15 mm) long, borne on short stalks and congested in a terminal raceme.

Fruits: Flattened, nonhairy pods with obvious constrictions between the seeds.

Ecology: Grows on rocky slopes and on gravel bars in rivers and streams, from lowlands to the alpine zone; widespread in the Arctic tundra, ranging southward into the mountains of northern British Columbia.

A similar species, **northern sweet-vetch** *(H. boreale)* also grows in northern British Columbia, but it is primarily a Rocky Mountain plant. It has somewhat larger flowers than alpine sweet-vetch, and the leaflets are not conspicuously veined. The two species grow in similar habitats.

Purple Peavine *Lathyrus nevadensis*

This is a perennial herb that spreads by rhizomes. The stems are angled and weak, clambering over associated vegetation.

Leaves: Alternate, pinnately compound with two to five pairs of ovate to elliptical leaflets and terminal tendrils.

Flowers: Showy, sweetpea-like, lavender, two to several borne on stalks from leaf axils.

Fruits: Hairless pods.

Ecology: Several varieties of this species have been recognized, and collectively they grow in a variety of habitats, from lowland forests and thickets to subalpine meadows, in both dry and moist sites; widely distributed, including the mountains of Washington and southern British Columbia.

Purple peavine is similar to other species of *Lathyrus* and the related genus *Vicia*. However, this is the only species that grows above mid-elevations in the Pacific Northwest mountains.

Alpine Sweet-Vetch ***Hedysarum alpinum***

Northern Sweet-Vetch ***Hedysarum boreale*** —George W. Douglas photo

Purple Peavine ***Lathyrus nevadensis***

PHLOX FAMILY
Polemoniaceae

The phlox family has stereotyped butterfly and moth flowers; that is, the flowers are modified for pollination by insects with extra long tongues. The petals are fused into a narrow, elongate tube with five spreading lobes. Insects land on the flat platform made by the spreading lobes and probe deeply into the tube for the concealed nectar. The stamens and stigmas are usually enclosed within the tube, facilitating pollination by the probing tongue. In many species, the area around the tube opening is not the same color as the lobes, resulting in a "bull's-eye" nectar guide. The five sepals too are fused into a tube around the petals, discouraging insects such as bees from chewing through the petal tube and stealing the nectar without pollinating the flower. The five stamens are fused to and thus derived from the inside of the petal tube. The ovary is superior and has three styles, a diagnostic characteristic of the family.

Spreading Phlox
Phlox diffusa

Spreading phlox is a perennial herb or subshrub that forms extensive mats from a branching, woody base.

Leaves: Opposite and linear, less than 1 inch (5 to 20 mm) long, densely overlapping with tufts of "wool" at the leaf bases; leaf tips sharp pointed but not spiny.

Flowers: Showy, pink to lavender or white, solitary at the tip of short, erect stems; petal lobes about ¼ inch (5 to 10 mm) long.

Fruits: Few-seeded capsules.

Ecology: A striking wildflower, nearly ubiquitous on rocky, alpine ridges in Washington and southern British Columbia; also grows on rock outcrops in high-elevation forests.

Phlox is a large genus, best known on the sagebrush steppe. Some of the arid-land species, such as **tufted phlox** *(P. caespitosa),* extend upward into the subalpine zone along dry, south-facing slopes. This phlox has tufted, erect, short stems with a single, relatively large white flower at the end. The leaves are narrowly linear, rigid, and spine tipped. Tufted phlox is an uncommon cushion plant in the mountains of eastern and central Washington. It prefers dry, rocky places, often growing among ponderosa pines.

Another steppe species that occasionally extends well up into the mountains on the eastern side of the Cascades in Washington and adjacent British Columbia is **long-leaf phlox** *(P. longifolia).* This species has showy, dark pink to lavender flowers and long, soft, linear leaves. It grows in dry, rocky sites, often with sagebrush.

Spreading Phlox *Phlox diffusa*

Tufted Phlox *Phlox caespitosa*

Long-Leaf Phlox *Phlox longifolia*

Showy Jacob's Ladder *Polemonium pulcherrimum*

This species, a perennial herb, has weak, mainly erect, clustered stems, often taller than 1 foot (30 cm).

Leaves: Alternate, pinnately compound, the many leaflets borne in pairs, looking like the rungs on a ladder, each leaflet about 1 inch (2 to 3 cm) long.

Flowers: Showy, as the common and Latin names suggest (*pulcherrimum* means "showy"), pale blue to lavender, with a yellow-orange "bull's-eye" ring (nectar guide) at the base of the petal lobes; flowers more or less bell shaped, clustered at the branch tips.

Fruits: Few-seeded capsules.

Ecology: Common plant of open forests and meadows at middle to high elevations; prefers rocky places and is distributed throughout the Pacific Northwest, although it is largely replaced by the more robust **tall Jacob's ladder** *(P. caeruleum)* in southeastern Alaska and northern British Columbia; tall Jacob's ladder has pointed petals.

Elegant Jacob's Ladder *Polemonium elegans*

Elegans means "elegant," or showy, which describes this low, more or less matted, perennial herb. Also called sky pilot, this species has clustered stems up to 6 inches (15 cm) high. Sticky hairs clothe the entire plant.

Leaves: Mainly basal and pinnately compound; leaflets numerous and crowded, about ¼ inch (3 to 6 mm) long.

Flowers: Showy, pale blue to lavender, with a yellow-orange "bull's-eye" at the base of the petal lobes; flowers trumpet shaped, loosely to more often densely clustered at the stem tips.

Fruits: Few-seeded capsules.

Ecology: Rocky slopes and rock crevices in the alpine zone of the Cascades in Washington and adjacent British Columbia.

Some of *Polemonium* species, including this one, typically have a slight to strong skunky odor produced by gland-tipped hairs. The odor reputedly repels herbivores and undesirable pollinators. The skunkiest of the species, appropriately, is called **skunky Jacob's ladder** *(P. viscosum).* It is a mat-forming plant with showy blue flowers that resemble those of elegant Jacob's ladder but are usually darker blue. It also has smaller, more numerous leaflets. Skunky Jacob's ladder is primarily a Rocky Mountain species, restricted to alpine, rocky habitats in the northeastern and north-central mountains of Washington and the Coast Mountains of British Columbia.

Showy Jacob's Ladder ***Polemonium pulcherrimum***

Elegant Jacob's Ladder ***Polemonium elegans***

Skunky Jacob's Ladder ***Polemonium viscosum***

PINK FAMILY Caryophyllaceae

The pinks comprise a large family, well represented in western North America. Based on ecological specialties, the family can be divided into three groups: dry-land species, weeds, and arctic-alpine species. Dry-land species are primarily annuals, escaping drought by completing their life cycle during moist periods. The weeds, predictably, are opportunists and take advantage of disturbed areas, such as cultivated fields and gardens. The arctic-alpine species are usually cushion perennials, an adaptation that protects plants from exposure in hostile, windy environments.

Species in this family can be recognized by a combination of vegetative and floral characteristics. The plants have opposite, undivided leaves borne on enlarged nodes. The flowers are radially symmetrical and floral parts come in fives–five petals, five sepals, and five or ten stamens. The ovary matures into a capsule with numerous seeds borne along a central column. With some exceptions, the petals are notched or divided at the tip.

Moss Campion *Silene acaulis*

Moss campion is one of the most beautiful plants in the alpine "rock gardens." It is *the* stereotype cushion plant, exemplifying an adaptation that protects the plants from drying, freezing winds while enabling them to take full advantage of available sunlight.

This species has an interesting and complex reproductive strategy. Many of the plants (cushions) have unisexual flowers–either male or female; other plants have bisexual flowers. This mix of sexual types provides the best of both worlds: it increases genetic variability and adaptiveness through crossbreeding among the unisexual plants and at the same time ensures the security of self-pollination for the bisexual types. Self-pollination is important in a hostile environment where pollinator activity is unpredictable.

Moss campion grows from a branched, woody root crown. The cushion is only a few inches high, but often exceeds 1 foot (30 cm) in diameter.

Leaves: Opposite, congested at the base of short, erect flowering stems, linear to lance shaped; about ⅛ to ½ inch (4 to 12 mm) long.

Flowers: Pink to lavender or rarely white, borne singly at the tip of 1- to 2-inch (2 to 5 cm) stems; petals about ⅜ inch (8 to 12 mm) long, notched at the tip, with two small appendages near mid-length; sepals fused into a bell-shaped tube.

Fruits: Many-seeded capsules.

Ecology: A common arctic-alpine species of western North America; found throughout the region on rocky ridges, scree slopes, rock faces, and in rock crevices.

Moss Campion *Silene acaulis*

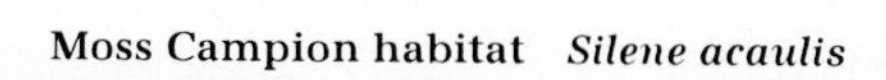

Moss Campion habitat *Silene acaulis*

Thread-Leaved Sandwort *Arenaria capillaris*

A plant of dry habitats, this perennial herb has clustered stems derived from a branched root crown. The stems are erect, up to 1 foot (30 cm) high, and covered with gland-tipped hairs near the top.

Leaves: Basal and opposite, linear (threadlike), 1 to 2 inches (2 to 5 cm) long.

Flowers: White, three to several in an open cluster at the stem tip; petals slightly notched at the tip, about ⅜ inch (1 cm) long; sepals egg shaped, rounded at the tip.

Fruits: Many-seeded capsules.

Ecology: A drought-tolerant species that ranges from the sagebrush steppe to the alpine zone, growing on open, rocky ridges throughout Washington and southern British Columbia.

Boreal Sandwort *Minuartia (Arenaria) rubella*

Boreal sandwort is a low, herbaceous cushion plant covered with gland-tipped hairs. The root crown branches give rise to numerous spreading stems. The flowering stems are erect, up to 4 inches (10 cm) high, and sometimes branched.

Leaves: Mainly basal, numerous and overlapping, linear to narrowly lance shaped, ⅛ to ⅜ inch (3 to 12 mm) long, three veined.

Flowers: White, few to several in open clusters; petals rounded at the tip, about ¼ inch (4 to 6 mm) long; sepals lance shaped, sharp pointed, three veined, and about the same length as the petals.

Fruits: Many-seeded capsules.

Ecology: A common cushion plant in moist, rocky places, often along stream banks and on wet scree slopes; distributed throughout the Pacific Northwest, usually in the alpine zone.

This species closely resembles arctic sandwort, but has somewhat smaller flowers, with sharp-pointed sepals (not blunt) and three-veined leaves (not one-veined). The sepals are usually reddish purple (*rubella* means "red").

Thread-Leaved Sandwort *Arenaria capillaris*

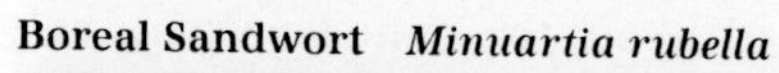

Boreal Sandwort *Minuartia rubella*

Boreal Sandwort *Minuartia rubella*

Arctic Sandwort *Minuartia (Arenaria) obtusiloba*

Gland-tipped hairs cover this low, mat-forming perennial herb. Its many spreading stems grow out of an extensively branched root crown. The flowering stems are erect and 1 to 2 inches (2 to 5 cm) high.

Leaves: Opposite to whorled, narrowly lance shaped to linear, about ⅜ inch (5 to 10 mm) long, with a single, prominent vein.

Flowers: White, solitary on flowering stems; petals rounded at the tip and about ¼ inch (6 to 8 mm) long; sepals three-veined, elongate, and rounded at the tip.

Fruits: Many-seeded capsules.

Ecology: An arctic-alpine species that ranges from Alaska through the Washington mountains; grows on rocky ridges in the alpine and upper subalpine zones, particularly on the drier, east-facing mountain slopes.

Several species of *Minuartia* mimic arctic sandwort in appearance and habitat preference, including boreal sandwort, and **Ross sandwort** *(M. austromontana).* Ross sandwort is a denser cushion plant and has much smaller flowers, with petals less than ⅛ inch (3 mm) long.

Bigleaf Sandwort *Mochringia (Arenaria) macrophylla*

Bigleaf sandwort is a low, perennial herb that spreads by rhizomes. The stems, 2 to 8 inches (5 to 20 cm) long, are weak and usually branched.

Leaves: Opposite, narrowly elliptical to lance shaped, 1 to 2 inches (2 to 6 cm) long and up to ⅝ inch (15 mm) wide; the tip is pointed.

Flowers: White, two to five in loose clusters; petals about ¼ inch (5 to 8 mm) long, rounded at the tip, and marked with green at the base; sepals sharp pointed, about the same length as the petals.

Fruits: Few-seeded capsules.

Ecology: A sandwort with a broad ecological tolerance, it grows in a variety of habitats from dry forests to moist meadows, often in rocky places; has a broad elevational range, too, from lowlands to subalpine meadows; distributed through Washington and southern British Columbia.

Blunt-leaf sandwort *(M. lateriflora)* closely resembles bigleaf sandwort, but it grows primarily in the northern Pacific Northwest, southeastern Alaska, and northern British Columbia. It grows in similar habitats as bigleaf sandwort, but has rounded sepals and leaf tips.

Arctic Sandwort *Minuartia obtusiloba*

Ross Sandwort *Minuartia austromontana*

Bigleaf Sandwort *Moehringia macrophylla*

Field Chickweed *Cerastium arvense*

Dense, white hair covers these perennial herbs. The plants often form mats, with numerous stems 2 to 12 inches (5 to 30 cm) long.

Leaves: Basal and opposite, linear to lance shaped, about 1 inch (3 cm) long.

Flowers: White, rather showy, two to five per stem; petals about ⅜ inch (1 cm) long, deeply notched at the tip; sepals about ¼ inch (4 to 6 mm) long.

Fruits: Many-seeded capsules.

Ecology: Highly variable, this species ranges from lowlands to alpine ridges, mainly on rock outcrops and other rocky places; widely distributed in Washington and British Columbia, but replaced in Alaska by the very similar **alpine chickweed** *(C. beeringianum),* which is rare in southern British Columbia and Washington.

Parry's Silene *Silene parryi*

Parry's silene is a perennial herb with several clustered stems about 1 foot (30 cm) high arising from a branched root crown. Gland-tipped hairs cover the plant.

Leaves: Basal and opposite, narrowly elliptical to linear, 1 to 3 inches (3 to 8 cm) long.

Flowers: Borne in clusters of three to seven at the stem tip and in leaf axils; sepals green to purplish, fused, bell shaped, conspicuously veined, and sticky; the five petals are white to purplish and divided at the tip into four segments.

Fruits: Many-seeded capsules.

Ecology: A plant primarily of dry rocky places, such as ridges and meadows, mainly below the alpine zone; distributed throughout Washington and southern British Columbia.

Long-Stalked Starwort *Stellaria longipes*

Stellaria means "star," and starwort means "starplant," thus the star shape of the flower is doubly noted. The plant is a low, perennial herb with solitary or clustered angled stems, 2 to 8 inches (5 to 20 cm) high. It spreads by rhizomes.

Leaves: Opposite, stiff, bluish green, narrowly lance shaped, with a sharp tip; ⅜ to 1¼ inches (1 to 3 cm) long.

Flowers: White, borne singly or a few at the stem tips; petals deeply divided, about ¼ inch (5 to 7 mm) long; sepals acute, three veined.

Fruits: Many-seeded capsules.

Ecology: Moist, rocky slopes and stream banks in the mountains throughout the Pacific Northwest.

Field Chickweed *Cerastium arvense*

Parry's Silene *Silene parryi*

Long-Stalked Starwort *Stellaria longipes*

Primrose Family Primulaceae

Although there is much variation among primrose family species, they share a combination of characteristics: flower parts come in fives—five sepals, five petals, and five stamens—and the flowers are radially symmetrical; the petals are fused into a tube with five lobes; the seeds of the ovary are arranged along a central column; and the leaves, with a few exceptions, are basal.

Fairy-Candelabra *Androsace septentrionalis*

This diminutive plant is one of the few annuals that grows in the alpine zone in western North America. Although it is small and easily overlooked, it is as charming as its common name suggests. It has many stems, 1 to 8 inches (3 to 20 cm) high, derived from a slender taproot.

Leaves: Basal, narrowly elliptical, ⅜ to 1¼ inches (1 to 3 cm) long, sometimes toothed.

Flowers: Small and individually inconspicuous, borne in umbels; sepals usually reddish; petals white, about ⅛ inch (3 to 5 mm) long.

Fruits: Several-seeded capsules.

Ecology: Distributed sporadically in the mountains throughout the Pacific Northwest, mainly in rocky, exposed sites.

Few-Flowered Shootingstar *Dodecatheon pulchellum*

The flowers of this species, as well as other species of *Dodecatheon,* fancifully resemble shooting stars. This widely distributed perennial herb grows with a solitary, unbranched stem.

Leaves: Basal, elliptical, 1 to 8 inches (3 to 20 cm) long, usually hairless and nontoothed.

Flowers: Showy, pale lavender to purple or occasionally white, one to several in a terminal umbel; five petal lobes trailing backward from a yellow base and the style, which is enclosed by dark purple anthers, project forward and downward.

Fruits: Many-seeded capsules.

Ecology: The most common shootingstar, distributed throughout the Pacific Northwest in moist to wet meadows and seepages, from lowlands into the alpine zone.

Jeffrey's shootingstar *(D. jeffreyi)* is also common in the region, growing in similar habitats as few-flowered shootingstar, but it is a taller plant with narrower, lance-shaped leaves.

Fairy-Candelabra *Androsace septentrionalis*

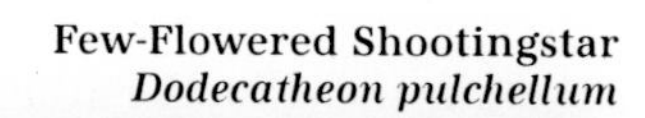

Few-Flowered Shootingstar
Dodecatheon pulchellum

Jeffrey's Shootingstar
Dodecatheon jeffreyi

Smooth Douglasia

Douglasia laevigata

When in bloom, this is a spectacular cushion plant. It is a dense, mat-forming perennial with prostrate branches that spread from a woody base. Minute, branched (star-shaped) hairs cover the stems.

Leaves: Concentrated at the base of branches and erect stems, more or less cylindrical, about ¼ to ¾ inch (5 to 20 mm) long, usually toothed.

Flowers: Dark pink to purplish, collectively very showy, two to several in tight umbels at the stem tips; petals about ⅜ inch (1 cm) long, including the tube and lobes.

Fruits: Few-seeded capsules.

Ecology: Found on rocky alpine ridges and talus slopes of the Cascades in Washington and Olympics; a rare species on Vancouver Island.

Arctic Starflower

Trientalis arctica

A small, bog-loving, perennial herb, this species propagates from small, underground tubers. The stems are solitary, erect, and 3 to 8 inches (7 to 20 cm) high.

Leaves: Whorled at the stem tip, often with a few reduced leaves below on the stem, egg shaped to elliptical, 1 to 2 inches (2 to 5 cm) long.

Flowers: White to pink tinged, star shaped, usually single on a threadlike stalk from the leaf whorl; petals five to seven.

Friuts: Few-seeded capsules.

Ecology: Moist to wet open forests and meadows, from lowlands up to the alpine zone, often in bogs; distributed throughout the Pacific Northwest but most common in Alaska and northern British Columbia.

A similar species, **woodland starflower** *(T. latifolia),* is common in lowland forests of Washington and southern British Columbia. It has larger leaves and more flowers (two to five) per plant.

Several primroses grow in the Arctic tundra of Alaska, and a few species reach southward. The most common in the Pacific Northwest is **wedge-leaf primrose** (*Primula cuneifolia* ssp. *saxifragifolia*), a delicate perennial herb with erect, solitary stems up to 6 inches (15 cm) high. Its leaves are basal and spoon shaped, and the tip is conspicuously toothed. The flowers are showy, dark pink to purplish, usually solitary, and about ¾ inch (2 cm) across. The petal tube is narrow, and the petal lobes are spreading and deeply heart shaped. It ranges southward into northern British Columbia and is rare on northwestern Vancouver Island. It prefers moist subalpine and alpine meadows.

Smooth Douglasia *Douglasia laevigata* —Rich Fonda photo

Arctic Starflower *Trientalis arctica*

Wedge-Leaf Primrose *Primula cuneifolia* —George W. Douglas photo

Purslane Family Portulacaceae

Species in this small family of succulent herbs have rubbery, usually smooth-margined leaves. Generally, the flowers have two sepals—an unusual number—five petals, which are usually pink striped, and five stamens. The stamens are borne at the base of the petals, opposite them, and usually slightly fused to them. This generality is violated by species of *Lewisia,* which have more petals and stamens. **Pussypaws** *(Spraguea umbellata),* which resembles buckwheat, is unusual in the family. It has prostrate, branched stems with a headlike umbel of small flowers at the ends. It grows in wet, gravelly sites in the high mountains of southern British Columbia and Washington.

Western Spring Beauty *Claytonia lanceolata*

This is indeed a spring beauty, conspicuous with an early spring appearance in the colorless landscape uncovered as the snow retreats. *Lanceolata* describes this perennial herb's lance-shaped leaves. This species is succulent, brittle stemmed, and hairless. The stems are solitary or few, up to 8 inches (20 cm) high, and derived from an underground bulb, as are the leaves.

Leaves: Basal and opposite; basal leaves one to several, with long stalks; opposite leaves two, situated near mid-length on the stem, nonstalked, lance shaped, and 1 to 3 inches (2 to 7 cm) long.

Flowers: White to pink or rarely yellow, usually marked with darker pink stripes, few to many in open racemes above the two leaves; petals about ¼ to ¾ inch (8 to 18 mm) long.

Fruits: Few-seeded capsules.

Ecology: A species with a broad ecological range, from the sagebrush steppe and forests on the eastern side of the mountain ranges to subalpine and alpine meadows; distributed throughout the mountains of Washington and southern British Columbia.

A yellow-flowered form of this variable species grows in the North Cascades of Washington. This color form, often placed in a separate variety (var. *chrysantha*), intergrades with the typical form and probably differs only in one or a few genes that affect flower color.

In northern British Columbia and southeastern Alaska, western spring beauty is replaced by the **Alaska spring beauty** *(C. sarmentosa),* which has larger flowers with red-striped, rosy petals. It grows from rhizomes rather than a bulb and often forms mats with numerous flowering stems, each with a pair of opposite leaves. It prefers moist to wet habitats, below snowfields and along streams, and ranges upward from lowlands into the subalpine zone.

Western Spring Beauty *Claytonia lanceolata*

Western Spring Beauty *Claytonia lanceolata*

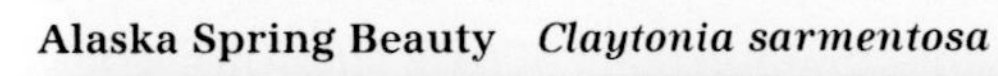

Alaska Spring Beauty *Claytonia sarmentosa*

Siberian Miner's Lettuce *Claytonia sibirica*

All species of miner's lettuce are renowned for their edibility, both the succulent stems and the leaves. This species is a succulent annual (sometimes perennial) hairless herb, with fibrous roots. The stems, usually several, are weak and sprawling, about 1 foot (20 to 40 cm) long.

Leaves: Basal and opposite; basal leaves several, broadly elliptical, with long petioles; opposite leaves two, egg shaped, stalkless, and situated near mid-length on the stem.

Flowers: White to pinkish, often striped, several in one to three racemes above the opposite leaves; petals ¼ to ½ inch (6 to 12 mm) long, notched at the tip.

Fruits: A capsule with one or a few seeds.

Ecology: A common species in moist forests and along streamsides, ranging from lowlands to the subalpine zone. Widely distributed in Washington and southern British Columbia.

Columbia Lewisia *Lewisia columbiana*

Attractive flowers mark this succulent, herbaceous perennial, which has one to several stems derived from a thick, fleshy taproot. The stems are branched and up to 1 foot (30 cm) high.

Leaves: Basal (the stem leaves are reduced and bractlike), numerous, linear or strap shaped, 1 to 5 inches (3 to 12 cm) long.

Flowers: White or pink with magenta stripes, a few to several on each of the branches; petals (seven to ten) about ¼ to ½ inch (5 to 12 mm) long; stamens five to seven.

Fruits: Few-seeded capsules.

Ecology: A locally common plant of rock outcrops, ledges, and talus slopes in the high mountains of Washington and southern British Columbia, particularly on the eastern slopes in the Cascades and Olympics.

Pygmy or **dwarf lewisia** *(L. pygmaea)* is a diminutive lewisia with a similar distribution. This charming little plant has linear basal leaves and a few to several short stems, 1 to 3 inches (2 to 8 cm) long, each bearing a single flower. The showy flowers resemble those of Columbia lewisia in size, color, and number of parts. However, the plants usually go unobserved because they are generally hidden among the associated vegetation. Pygmy lewisia is predominantly an alpine species.

Siberian Miner's Lettuce *Claytonia sibirica*

Columbia Lewisia *Lewisia columbiana*

Dwarf Lewisia *Lewisia pygmaea*

ROSE FAMILY Rosaceae

The extensive variation in this large family makes difficult the task of writing a simple description that fits all members while excluding species of other families. The family includes trees and shrubs as well as annual and perennial herbs, and leaves among the species vary from extensively divided to simple. Most species in the family have flowers that are large and showy, but in others they are small and inconspicuous, even incomplete. Nonetheless, the flowers provide the only consistent link among the diverse species. With few exceptions, the flowers have five petals, five sepals, and numerous stamens. Most important, these floral parts are borne not on the receptacle but on a cup-shaped, saucer-shaped, or tubular structure called a calyx tube or hypanthium. The hypanthium usually surrounds the single to numerous pistils, but in one group, the ovary is located below the flower parts (inferior), and the hypanthium is positioned on top of the ovary. In any case, the hypanthium is not always obvious.

Floral configuration in the rose family is regarded as primitive, especially the multiple pistils and numerous stamens. The flowers exhibit an unspecialized pollination strategy, with the nectar and pollen accessible to all types of foraging insects. Ecologically, however, the importance of the family cannot be overemphasized. Many species are dominant members of their community and many serve as a valuable food source, especially the ones that produce fleshy fruits.

Serviceberry *Amelanchier alnifolia*

Serviceberry is an important winter browse plant for deer and elk, especially on open, relatively snow-free ridges. The berries are important for birds and small mammals, too. This large deciduous shrub or small tree spreads horizontally by short rhizomes, often forming dense thickets.

Leaves: Alternate and stalked; blades broadly elliptical to circular, rounded at the base, and toothed, at least toward the tip, 1 to 2 inches (2 to 4 cm) long.

Flowers: White, showy and fragrant, borne in clusters of a few to many on short branches; petals five, narrowly elliptical, about ½ inch (1 to 2 cm) long; ovary inferior; styles five.

Fruits: Dark blue to purple edible berry; sepals persist on the mature berry.

Ecology: Although this species grows in coniferous and deciduous forests, particularly on the western side of the Cascades in Washington and Coast Mountains of British Columbia, it is most conspicuous on open hillsides and along canyon walls; extends upward from lowlands into the subalpine zone and is distributed throughout the Pacific Northwest.

Serviceberry flowers *Amelanchier alnifolia*

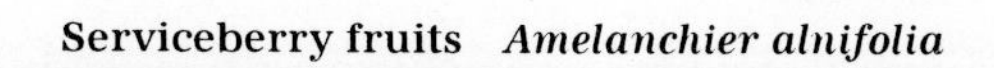

Serviceberry fruits *Amelanchier alnifolia*

Goat's Beard *Aruncus dioecus*

A robust, hairless, perennial herb, up to 6 feet (2 m) tall, that spreads by rhizomes. The plants usually have several stems.

Leaves: Alternate, lower leaves very large, divided, and spraylike (usually three times compound), upper leaves smaller, leaflets sharp pointed and toothed.

Flowers: White, very small but densely clustered along the ends of branches, unisexual—male and female flowers borne on separate plants.

Fruits: Small, few-seeded pods.

Ecology: A species of moist forests and clearings, such as avalanche tracks; most conspicuous along streams and roadsides; ranges from lowlands upward into subalpine meadows and grows throughout the region.

A large, deciduous shrub that has equally small but bisexual white flowers is **ocean spray** *(Holodiscus discolor).* In this species, though, the flowers are more congested. Also, the leaves are simple, shallowly lobed, somewhat triangular, and prominently veined. Ocean spray grows in drier sites than goat's beard, preferring forested and open sites from lowlands to mid-elevations; it is distributed from southern British Columbia through Washington.

Mountain Avens *Dryas octopetala*

This is a dwarf, prostrate, evergreen shrub with spreading branches that form extensive mats. The erect flowering stems are only 1 to 6 inches (3 to 15 cm) high.

Leaves: Alternate on prostrate branches, leathery, and lance shaped, ⅜ to 1¼ inches (1 to 3 cm) long; leaf surface is raised between veins, and the leaf margins are coarsely toothed and rolled under.

Flowers: Large and showy, white to cream colored, solitary on erect stems; petals usually eight, about ⅜ inch (1 cm) long; hypanthium saucer shaped.

Fruits: Numerous one-seeded achenes with long, feathery styles.

Ecology: Rocky sites in the mountains, usually on alpine ridges along the eastern side of the Cascade crest in Washington and the mountains of British Columbia.

D. integrifolia is a more common species in Alaska and northern British Columbia. It differs in having smaller, less prominently toothed leaves. The two dryads have overlapping ranges and habitats, and apparently hybridize.

Goat's Beard *Aruncus dioecus*

Ocean Spray *Holodiscus discolor*

Mountain Avens *Dryas octopetala*

Wild Strawberry *Fragaria vesca*

This is a low, perennial herb that spreads by stolons (runners). Its flowering stems are erect and up to 4 inches (10 cm) high.

Leaves: Primarily basal with long petioles, compound, with three leaflets, leaflets toothed and conspicuously veined, area between the veins is raised.

Flowers: White, showy, solitary or few on erect stems; petals about ⅜ inch (8 to 10 mm) long, broadly egg shaped; hypanthium saucer shaped.

Fruits: Fleshy and berrylike, formed from the fleshy receptacle and numerous one-seeded pistils; delicious.

Ecology: Common in the mountains of Washington and southern British Columbia, typically in areas with sparse vegetation such as roadsides and trails; also grows in moist open forests, along stream banks, and in meadows.

Some plant species in other genera have strawberry-like leaves and may be confused with strawberries when not in flower or fruit. These include some of the cinquefoils and **sibbaldia** *(Sibbaldia procumbens),* which is a small, cushion perennial that spreads by rhizomes. The flowers are small and inconspicuous, with pale yellow petals that are shorter than the sepals. The flowers have a set of five bracts below and are similar to but smaller than the sepals. This species is an unusual representative of the rose family because its flowers have fewer than ten stamens. Though infrequently observed, it grows on alpine ridges throughout the Pacific Northwest.

Alpine Avens *Geum rossii*

Alpine avens is another attractive alpine wildflower that spreads by rhizomes and forms large, dense mats. The flowering stems are erect, unbranched, and up to 8 inches (20 cm) high.

Leaves: Alternate and basal, 2 to 4 inches (4 to 10 cm) long, pinnately compound, the leaflets further lobed at the tip.

Flowers: Showy, bright yellow, one to three on flowering stems; petals about ½ inch (8 to 12 mm) long; hypanthium bowl shaped and surrounded by five sepal-like bracts.

Fruits: Numerous one-seeded achenes, each with a persistent, spinelike style.

Ecology: Common in the Arctic tundra, Aleutian Islands, and the alpine zone of the Rocky Mountains; it extends into southeastern Alaska and grows on the eastern side of the Cascade crest in Washington. It prefers rocky, alpine ridges.

Wild Strawberry *Fragaria vesca*

Sibbaldia *Sibbaldia procumbens*

Alpine Avens *Geum rossii*

Indian Plum
Oemleria cerasiformis

A medium-sized to large deciduous shrub, it grows 6 to 13 feet (2 to 4 m) high and has smooth, purplish bark.

Leaves: Alternate, blades symmetrically elliptical, nontoothed, 2 to 5 inches (5 to 12 cm) long; somewhat paler on the undersurface.

Flowers: White to greenish, unisexual (male and female flowers borne on separate plants); the flowers hang in pendant racemes from leaf axils; petals about ¼ inch (5 to 8 mm) long; hypanthium bell shaped.

Fruits: Purple drupes (like plums); edible but bitter.

Ecology: A very common shrub in lowland to mid-elevation forests on the western side of the Cascades in Washington, extending northward onto Vancouver Island.

Indian plum flowers early in spring, before the leaves come out on deciduous trees. Covered with flowers and growing in abundance, Indian plum is a welcome harbinger of spring.

Bitter Cherry
Prunus emarginata

A large shrub to small tree; bitter cherry has purplish bark that is indeed bitter, especially on young twigs.

Leaves: Alternate, blades elliptical, rounded at the tip, and finely toothed, 1 to 3 inches (3 to 8 cm) long; petioles nearly as long.

Flowers: White, borne in clusters of a few to several along the twigs; petals about ¼ inch (5 to 7 mm) long; the hypanthium is tubular and encloses the ovary.

Fruits: Dark red to purple drupes (like cherries).

Ecology: Moist forests to open hillsides, from low to mid-elevations in the mountains of Washington and southern British Columbia.

Generally, bitter cherry is divided into two varieties. Variety *mollis* is a tree form found on the western side of the Cascades and coastal ranges of British Columbia. Variety *emarginata* is more shrubby and drought tolerant, and grows primarily on the drier, eastern side of the mountains.

Chokecherry *(P. virginiana)* is a large shrub or small tree that also grows on the eastern side of the mountains, often reaching high elevations. It differs from bitter cherry in having dense, elongate, pendant clusters of flowers and fruit. The fruit makes excellent syrup and jellies.

Indian Plum flowers
Oemleria cerasiformis

Indian Plum fruits
Oemleria cerasiformis

Bitter Cherry *Prunus emarginata*

Chokecherry *Prunus virginiana*

Partridge Foot *Luetkea pectinata*

A common high-altitude perennial herb with stems that are somewhat woody, partridge foot spreads by rhizomes and stolons, forming expansive mats. The stems are erect and 3 to 6 inches (8 to 15 cm) high.

Leaves: Mainly basal, deeply lobed at the tip, resembling the toes of a partridge foot, up to 1 inch (2 cm) long, including the petiole.

Flowers: White, or yellowish because of the numerous stamens, borne in a congested raceme at the stem tip; petals about ⅛ inch (3 to 4 mm) long; hypanthium bowl shaped.

Fruits: Four or five several-seeded pods.

Ecology: A common, attractive plant found in moist meadows and on rocky slopes in the subalpine and alpine zones, especially in areas with deep snow and late snowmelt; often grows within and around tree clumps and is distributed throughout the Pacific Northwest.

Shrubby Cinquefoil *Potentilla fruticosa*

A prostrate to erect, extensively branched, deciduous shrub, this cinquefoil has reddish brown, typically shredding bark.

Leaves: Alternate, pinnately compound, usually with five narrowly elliptical, nontoothed leaflets, each ⅜ to ¾ inch (1 to 2 cm) long and covered with silky hairs.

Flowers: Bright yellow, showy, about 1 inch (2.5 cm) across, solitary on short branches; hypanthium saucer shaped and surrounded by five sepal-like bracts.

Fruits: Numerous single-seeded achenes covered with silky hairs.

Ecology: A variable species that grows in a number of habitats, from meadows in the sagebrush steppe to rocky alpine ridges; inhabits bogs and muskegs in the Arctic; often a dominant species in the subalpine zone in northern British Columbia, and is primarily a mat-forming alpine plant in southern British Columbia and Washington.

Several *Potentilla* species grow in the Pacific Northwest mountains, mainly in the alpine or subalpine zone. Although none can match shrubby cinquefoil for sheer beauty, some species are dominant plants and put on a colorful display. A few are diminutive cushion plants, such as **snow cinquefoil** *(P. nivea),* which is a densely hairy plant with trifoliate leaves, strawberry-like. It is widespread in the alpine zone, but seldom abundant.

Partridge Foot *Luetkea pectinata*

Shrubby Cinquefoil *Potentilla fruticosa*

Shrubby Cinquefoil *Potentilla fruticosa*

Diverse-Leaved Cinquefoil *Potentilla diversifolia*

The most common alpine cinquefoil, this perennial herb has a few to several erect to spreading, branched stems. It grows about 1 foot (30 cm) high.

Leaves: Mainly basal, palmately compound; leaflets five—deeply toothed, with silky hairs beneath; petioles longer than the divided leaf blade.

Flowers: Bright yellow, about ¾ inch (2 cm) across, scattered along the branched stem; hypanthium saucer shaped and surrounded by five sepal-like bracts.

Fruits: Numerous one-seeded achenes.

Ecology: Meadows and rocky slopes throughout the Pacific Northwest.

Fan-Leaf Cinquefoil *Potentilla flabellifolia*

A perennial herb with a branched root crown, this cinquefoil often forms large clumps of erect to spreading stems 6 to 12 inches (15 to 30 cm) long.

Leaves: Mainly basal, trifoliate (like a strawberry leaf), with long petioles; the three leaflets roundish and deeply toothed.

Flowers: Bright yellow; about 1 inch (2.5 cm) across, two to five at the tip of erect stems; hypanthium saucer shaped and surrounded by five lobed, sepal-like bracts.

Fruits: Numerous one-seeded achenes.

Ecology: A regular inhabitant of subalpine meadows and stream banks throughout the mountains of Washington and southern British Columbia.

Villous Cinquefoil *Potentilla villosa*

No more than 6 inches (15 cm) high, this is a low, cushionlike, perennial herb. Silky gray hairs clothe the plant.

Leaves: Mainly basal, cloverlike (trifoliate), with long petioles; the three leaflets deeply toothed.

Flowers: Bright yellow, about 1 inch (2.5 cm) across; hypanthium saucer shaped, surrounded by five sepal-like bracts.

Fruits: Numerous one-seeded achenes.

Ecology: Primarily an alpine plant partial to rocky areas—rock outcrops, ledges, and talus slopes; distributed throughout the region but seldom common.

Diverse-Leaved Cinquefoil *Potentilla diversifolia*

Fan-Leaf Cinquefoil *Potentilla flabellifolia*

Villous Cinquefoil *Potentilla villosa*

Nootka Rose *Rosa nutkana*

A medium-sized shrub, 3 to 6 feet (1 to 2 m) tall, nootka rose has branched stems with paired, stout spines at the leaf nodes.

Leaves: Alternate, pinnately compound with five to seven rounded to broadly elliptical toothed leaflets.

Flowers: Light pink to dark rose, large and showy, 2 to 4 inches (5 to 10 cm) across.

Fruits: Rose hips, formed by the enclosure of numerous achenes inside a red-orange leathery shell.

Ecology: A common species in open forests and thickets in the foothills of the western Cascades of Washington and the Coast Mountains of British Columbia and southeastern Alaska.

Prickly rose *(R. acicularis)* is a much more widespread rose in British Columbia and Alaska. It differs from the nootka rose by having prickles scattered over the stems, not just at the leaf nodes. It primarily inhabits lowlands but may extend upward to high elevations in the mountains.

Western Mountain Ash *Sorbus scopulina*

This is a large, deciduous shrub with several erect stems, which may grow to be as much as 16 feet (5 m) tall.

Leaves: Alternate, pinnately compound; leaflets nine to seventeen, narrowly elliptical, toothed along the full length, sharp pointed, 1 to 3 inches (2.5 to 7.5 cm) long.

Flowers: White, small and densely congested in large round or flat-topped clusters; petals about ¼ inch (4 to 6 mm) long; hypanthium borne on the ovary, which is inferior.

Fruits: Red-orange berries, edible but bitter.

Ecology: Common in open forests and avalanche tracks; occasionally grows in subalpine meadows; distributed throughout the region.

This variable species is often divided into two varieties. Also, it overlaps in distribution with **Sitka mountain ash** *(S. sitchensis)*, which has almost the same growth form and occupies similar habitats. Sitka mountain ash can be distinguished by its shorter leaflets, which are rounded at the tip and toothed only from the tip to near mid-length.

Nootka Rose *Rosa nutkana*

Western Mountain Ash flowers *Sorbus scopulina*

Western Mountain Ash fruits *Sorbus scopulina*

Nagoonberry *Rubus arcticus* ssp. *acaulis*

Nagoonberry is a low, perennial herb that spreads by woody rhizomes. The flowering stems are erect and 2 to 5 inches (4 to 12 cm) high.

Leaves: Alternate, two to three per stem, trifoliate (strawberry-like) or merely deeply three lobed (not divided to the midvein), lobes or leaflets irregularly toothed.

Flowers: Pink to red, showy, about 1 inch (2 to 3 cm) across, one or two on erect stems; petals five or six, about ½ inch (10 to 16 mm) long; hypanthium bowl shaped.

Fruits: Small cluster of red drupelets, like raspberries; palatable.

Ecology: Common in Alaska in wet meadows, bogs, muskegs, and along streams; it extends southward through the mountains of British Columbia, ranging from lowlands to the subalpine zone.

Cloudberry *Rubus chamaemorus*

This is a low, perennial herb that spreads by woody rhizomes. The flowering stems are erect and 5 to 8 inches (12 to 20 cm) high.

Leaves: Mainly basal, few, three to five lobed and irregularly toothed, roundish in outline and conspicuously veined.

Flowers: White, unisexual (male and female flowers borne on separate plants), solitary on erect stems; petals about ⅜ inch (10 mm) long; hypanthium bowl shaped.

Fruits: Small cloudlike cluster of fleshy drupelets, pale orange when mature.

Ecology: Abundant throughout much of Alaska in bogs and muskegs and southward into British Columbia, where it ranges from lowlands to mid-elevations in the mountains.

The range of cloudberry extends into northern Europe, where it has long been prized for its attractively colored and tasty "berries."

Dwarf bramble *(R. lasiococcus)* superficially resembles cloudberry. It has trailing stems that root at the nodes, and its leaves are roundish and deeply three lobed. The flowers are white but bisexual, and the cluster of fleshy drupelets is red. Common in the southern Cascades of Washington, it is less so in the northern Cascades and adjacent British Columbia. It grows in conifer forests from mid-elevations up to the treeline.

Nagoonberry ***Rubus arcticus***

Cloudberry ***Rubus chamaemorus***

Dwarf Bramble ***Rubus lasiococcus***

Strawberry Bramble *Rubus pedatus*

Strawberry bramble is often an understory dominant species, especially in the forests of the North Cascades. Its tasty "berries" provide food for birds and small mammals. A low, perennial herb that spreads by runners (stolons), this plant often forms dense mats.

Leaves: Basal or nearly so, palmately compound with five, irregularly toothed leaflets; petioles long and wiry.

Flowers: White, solitary on 1 to 2½ inch (2 to 6 cm), leafless stalks; petals about ½ inch (10 to 12 mm) long; hypanthium saucer shaped.

Fruits: Two to five fleshy, bright red drupelets; palatable.

Ecology: A common plant in moist conifer forests from mid-elevations upward to the treeline; also common in open areas with sparse vegetation, particularly where vegetation is repressed by late snowmelt; found throughout the Pacific Northwest.

Wild Blackberry *Rubus ursinus*

Several blackberry species thrive in lowlands on the western side of the Cascades, but this is the only native species. Although the plant is a serious nuisance, often forming impenetrable tangles, its fruit is truly delicious. It is a trailing, spiny shrub with long spindly stems that scramble over associated vegetation and root at the tip.

Leaves: Alternate, cloverlike (trifoliate), the three leaflets irregularly toothed, the terminal leaflet the largest; leaf axis and petiole spiny.

Flowers: White, unisexual (male and female flowers borne on separate plants); petals ½ to ¾ inch (12 to 18 mm) long; sepals and hypanthium covered by white, woolly hairs; three to several flowers, grouped at the tip of erect 8- to 12-inch (20 to 30 cm) branches.

Fruits: Dark purple cone-shaped cluster of fleshy drupelets.

Ecology: A common plant in lowland forests of the western Cascades in Washington and the Coast Mountains of British Columbia, extending upward to mid-elevations; especially abundant and bothersome in forest openings and logged areas.

Strawberry Bramble flower ***Rubus pedatus***

Strawberry Bramble fruits ***Rubus pedatus***

Wild Blackberry ***Rubus ursinus***

Thimbleberry

Rubus parviflorus

The Latin *parviflorus* is a misnomer; it means "small flower," but thimbleberry has larger flowers than any other *Rubus* species, at least in the Pacific Northwest. It is a medium-sized, nonspiny deciduous shrub that spreads by rhizomes. The stems are erect and up to 10 feet (2.5 m) tall.

Leaves: Alternate, palmately lobed (like a maple leaf), blades up to 8 inches (20 cm) long and as wide, sharply toothed; petioles also 8 inches long.

Flowers: White, showy, about 2 inches (5 cm) across, few to several at the stem tips; hypanthium saucer shaped.

Fruits: A thimblelike aggregate of fleshy drupelets, crimson, edible–with good flavor.

Ecology: A common plant of open forests and clearings, extending upward into subalpine meadows; particularly abundant along streams in the mountains; grows in the mountains of Washington and southern British Columbia.

Salmonberry

Rubus spectabilis

Salmonberry is an attractive shrub in all stages of its life cycle, but its aggressive growth pattern and typical spininess make it a bit of a nuisance. It is a medium-sized shrub, up to 13 feet (4 m) tall, and spreads aggressively by rhizomes. The stems are erect or arching, usually spiny, reddish brown, and typically assume a zigzag form.

Leaves: Alternate, trifoliate, irregularly toothed, the terminal leaflet the largest.

Flowers: Reddish purple, showy, about 1 inch (3 cm) across, solitary or few on short, leafy branches; hypanthium bowl shaped.

Fruits: Salmon to dark red aggregate of fleshy drupelets, like a raspberry; bland but edible.

Ecology: A common shrub in open forests and clearings, along streams, in avalanche tracks, and in logged areas, often forming dense thickets; also found in subalpine meadows; distribution extends along the western slopes of the Cascades and Olympics into the Coast Mountains of British Columbia, ranging northward into southeastern Alaska.

Thimbleberry ***Rubus parviflorus***

Salmonberry flowers ***Rubus spectabilis***

Salmonberry fruits ***Rubus spectabilis***

Birch-Leaved Spirea

Spiraea betulifolia

This small, rhizomatous shrub grows to be 40 inches (1 m) tall.

Leaves: Alternate, egg shaped to elliptical, irregularly toothed, 1 to 3 inches (2 to 8 cm) long, resembling birch leaves.

Flowers: White, small but densely concentrated in a flat-topped, branched inflorescence; hypanthium bowl shaped.

Fruits: Small, few-seeded pods, usually five per flower.

Ecology: Common on the eastern, drier slopes of the Cascades and Coast Mountains of British Columbia, growing in forests and clearings from low to high elevations.

In Alaska and northern British Columbia, this species is replaced by the similar **Steven's spirea** *(S. stevenii),* which has smaller, more numerous flower clusters and smaller leaves. It grows in open forests and meadows.

Subalpine Spirea

Spiraea densiflora

Densiflora means "dense" or "close flowered," which describes this plant's flower cluster. It is a small, extensively branched shrub, up to 40 inches (1 m) tall, and spreads by rhizomes. The bark is reddish brown.

Leaves: Alternate, egg shaped to elliptical, 1 to 2 inches (2 to 4 cm) long, coarsely toothed.

Flowers: Rose colored, very small but densely concentrated in several flat-topped or rounded inflorescences; pink stamens extend beyond the petals, adding to the flower's beauty; hypanthium bowl shaped.

Fruits: Few-seeded pods, four or five per flower.

Ecology: A fairly common plant of forest clearings, avalanche tracks, and open subalpine forests, from mid-elevations to treeline in the mountains of Washington and southern British Columbia.

Douglas spirea *(S. douglasii),* a similar but larger shrub, occasionally grows in moist meadows in the mountains but most frequently occupies wet sites in lowlands, particularly along streams and lakeshores. It differs from subalpine spirea by having flower clusters that are long and narrow. When the two species overlap in distribution, they hybridize and the distinction is lost. Douglas spirea also hybridizes with birch-leaved spirea, and the resulting hybrid is reputedly the source of the species ***Spiraea pyramidata.*** Douglas spirea is highly variable, with some variants growing more or less throughout Washington and British Columbia, barely reaching southeastern Alaska.

Birch-Leaved Spirea *Spiraea betulifolia*

Subalpine Spirea *Spiraea densiflora*

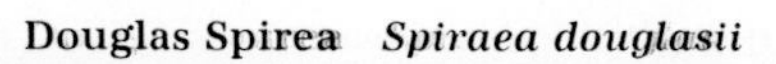

Douglas Spirea *Spiraea douglasii*

Sitka Burnet *Sanguisorba (sitchensis) canadensis* ssp. *latifolia*

This is another perennial herb that spreads by rhizomes. The stems are solitary or clumped, and grow to 40 inches (1 m) tall.

Leaves: Mainly basal, much reduced on flowering stems, pinnately compound, with nine to seventeen leaflets; the leaflets are 1 to 2 ½ inches (2 to 6 cm) long, elliptical, sharply toothed, and blunt tipped.

Flowers: White, very small and densely congested in terminal spikes; the stamens extend beyond the petals, making the flower cluster look like a bottlebrush.

Fruits: One-seeded achenes, one or two per flower.

Ecology: A boreal species, growing more or less throughout Alaska and northern British Columbia; prefers moist to wet habitats, such as bogs, muskegs, moist subalpine meadows, and stream banks. In Washington, it is restricted to the western side of the Cascades and Olympics, primarily at low elevations.

Sandalwood Family Santalaceae

A small family of perennial herbs, the sandalwoods all have alternate leaves. The flowers are small and inconspicuous, with five greenish purple sepals and no petals. The ovary is borne below the other flower parts (inferior), maturing into a juicy, usually red berry. The species are parasitic on the roots of other plants.

Geocaulon *Geocaulon livida*

Geocaulon spreads by rhizomes and has stems that are either solitary or clumped, 4 to 10 inches (10 to 25 cm) high. This species is sometimes placed in the genus *Comandra.*

Leaves: Alternate, thin and papery, egg shaped to elliptical, blunt tipped, 1 to 2 inches (2 to 5 cm) long.

Flowers: Small, greenish, and inconspicuous, usually in groups of three in leaf axils; petals lacking.

Fruits: Red, juicy, edible berries.

Ecology: A common plant of bogs and moist open woods in Alaska and northern British Columbia, extending southward to northern Washington.

Sitka Burnet *Sanguisorba canadensis*

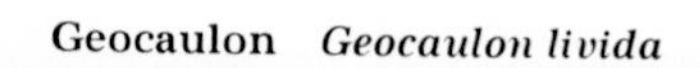

Geocaulon *Geocaulon livida*

Saxifrage Family   Saxifragaceae

This large family of perennial herbaceous plants has mainly basal leaves, and the leaves are usually palmately lobed or divided, like a maple leaf. The flowers are usually white, occasionally yellow or reddish, and are produced in open or dense inflorescences at the stem tips. The flower parts are mainly in fives—five sepals and petals and five or ten stamens. Like those of the rose family, saxifrage flowers usually have more than one pistil, and the sepals, petals, and stamens are borne on a saucer- or cup-shaped hypanthium. However, flowers of the rose family typically have more than ten stamens.

Leatherleaf Saxifrage   *Leptarrhena pyrolifolia*

Leatherleaf saxifrage spreads by rhizomes and has stems that are unbranched, covered with gland-tipped hairs, and 8 to 16 inches (20 to 40 cm) tall.

Leaves: Basal, shiny and leathery, elliptical; blades 1 to 4 inches (2 to 10 cm) long, toothed.

Flowers: White, very small in dense, roundish clusters at the tip of red stems.

Fruits: Two red, many-seeded pods.

Ecology: A common, often dominant plant in wet areas such as seepages and streamsides; it ranges upward from mid-elevations into the subalpine zone and grows throughout the Pacific Northwest.

Brewer's Miterwort   *Mitella breweri*

This member of the saxifrage family spreads by threadlike rhizomes and has stems that are unbranched, solitary or few, and 4 to 12 inches (10 to 30 cm) high.

Leaves: Basal, circular to heart shaped, irregularly toothed; blades 2 to 3 inches (4 to 8 cm) long and broad.

Flowers: Greenish, small, and inconspicuous in narrow, elongate racemes; petals divided into lateral, threadlike lobes like a comb and are borne with the five stamens on a green, saucer-shaped hypanthium.

Fruits: A two-lobed pod with numerous seeds.

Ecology: An inhabitant of moist, open forests and meadows from mid-elevations into the subalpine zone, throughout Washington and in southern British Columbia.

Several species of *Mitella* occur in similar habitats within the region. The species are often difficult to differentiate.

Leatherleaf Saxifrage fruits
Leptarrhena pyrolifolia

Brewer's Miterwort
Mitella breweri

Leatherleaf Saxifrage flowers *Leptarrhena pyrolifolia*

Fringed Grass-of-Parnassus

Parnassia fimbriata

This is a perennial herb with short rhizomes. The stems are solitary or clustered and stand 6 to 12 inches (15 to 30 cm) high.

Leaves: Basal except for one reduced leaf midway on the stem; blades shiny, bright green, strongly veined, broadly heart shaped; 1 to 2 inches (2 to 4 cm) long and as broad.

Flowers: White, solitary at the stem tip; petals about ⅜ inch (8 to 12 mm) long, the lower half bearing numerous thick and tangled white hairs; in addition to five fertile stamens, the flowers have sterile stamens modified into greenish yellow succulent disks with several lobes, each with a bulbous tip—disks borne at the petal bases.

Fruits: Papery capsule with numerous seeds.

Ecology: A common plant of wet meadows, seepages, and streamsides, ranging from mid-elevations up to the alpine zone; distributed throughout the region.

In southeastern Alaska and British Columbia, this attractive species is largely replaced by **northern grass-of-Parnassus** *(P. palustris),* which differs in having petals that are not fringed and leaves that are ovate rather than heart shaped. The two species share similar habitats.

Tufted Saxifrage

Saxifraga caespitosa

This small, mat-forming saxifrage has several erect stems and grows to be 6 inches (3 to 15 cm) high.

Leaves: Basal, stem leaves scalelike; blades usually densely hairy, divided at the tip into three terminal lobes.

Flowers: White, solitary or few at the stem tips; petals ¼ to ½ inch (5 to 12 mm) long.

Fruits: Two-lobed, many-seeded capsules.

Ecology: A common species in the Arctic tundra, following mountain ranges southward into Washington on rocky slopes and rock outcrops, in rock crevices, and on cliff faces.

Tufted saxifrage is a variable species, particularly in flower size, petal shape, stem height, and hairiness of the leaves. The plants vary in response to their environment, too; exposed alpine forms are lower and more matted.

Fringed Grass-of-Parnassus *Parnassia fimbriata*

Northern Grass-of-Parnassus
Parnassia palustris

Tufted Saxifrage
Saxifraga caespitosa

Spotted Saxifrage *Saxifraga bronchialis*

Spotted flowers distinguish this saxifrage. It grows in dense cushions, the numerous erect flowering stems 2 to 6 inches (5 to 15 cm) high. Spotted saxifrage has an interesting growth form; as the cushion expands, the older part of the cushion eventually dies, leading to fragmentation and formation of "new" clones.

Leaves: Basal and alternate, rigid but rather succulent, linear to lance shaped, spine tipped, ⅛ to ⅝ inch (5 to 15 mm) long; leaf margins have miniature spiny teeth.

Flowers: Small and showy; few to several in open, flat-topped clusters; petals ⅛ to ¼ inch (5 to 8 mm) long, white with reddish purple spots above the middle and yellow-orange spots at the base.

Fruits: Several-seeded capsules, two-parted at the tip.

Ecology: A common species on rocky ridgetops, in rock crevices, and scree slopes, mainly in the alpine zone but occasionally at lower elevations, especially in northern British Columbia and Alaska; distributed throughout the region.

A related species in northern British Columbia and southeastern Alaska is **three-toothed saxifrage** *(S. tricuspidata),* which resembles spotted saxifrage in growth form, habitat, and floral coloration but differs in having leaves with three prominent teeth at the tip, as the Latin and common names suggest.

Western Saxifrage *Saxifraga occidentalis*

This common saxifrage is a perennial herb with short rhizomes. The stems are solitary or few, strictly erect, and stand 6 to 15 inches (15 to 38 cm) high. Red glands are usually visible below the inflorescence.

Leaves: Basal; leaf blades elliptical to egg shaped, toothed, about 2 inches (5 cm) long, with longer petioles.

Flowers: Small, numerous in an open or dense, often pyramidal inflorescence at the stem tip; petals white to pinkish, about ⅛ inch (2 to 4 mm) long with two yellow spots.

Fruits: Deeply two-lobed, many-seeded capsules.

Ecology: An inhabitant of moist subalpine or alpine meadows and rocky seepage areas of Washington and southern British Columbia; extremely variable, but most varieties grow outside the Pacific Northwest.

Spotted Saxifrage
Saxifraga bronchialis

Western Saxifrage
Saxifraga occidentalis

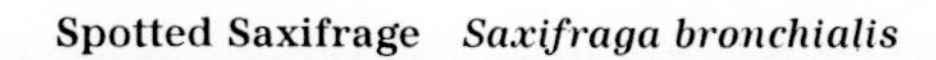

Spotted Saxifrage *Saxifraga bronchialis*

Rusty Saxifrage *Saxifraga ferruginea*

Rusty saxifrage has fibrous roots instead of rhizomes. The stems typically are branched, vary in height from a few inches to more than 1 foot (30 cm), and are usually sticky from gland-tipped hairs. Many or most of the flowers may be replaced by bulblets, which eventually fall to the ground and produce new plants. Nodding saxifrage exhibits the same characteristic.

Leaves: Basal, broadest toward the tip (club shaped), up to 5 inches (12 cm) long and 1 inch (2.5 cm) wide, irregularly toothed.

Flowers: Somewhat showy, a few to several on spreading branches; petals white to purplish, about ¼ inch (4 to 7 mm) long.

Fruits: Red, two-lobed, many-seeded capsules.

Ecology: A variable species found mainly in rocky but moist sites, such as along stream banks; ranges from lowlands to the alpine zone; distributed throughout the region.

Brook Saxifrage *Saxifraga nelsoniana (punctata)*

Named for its favorite habitat, this saxifrage spreads by short rhizomes. The stems are mainly unbranched, 6 to 18 inches (15 to 46 cm) tall, and hairy below the inflorescence.

Leaves: Basal, with long petioles and broadly heart shaped, coarsely toothed blades 1 to 2½ inches (2 to 6 cm) broad.

Flowers: White, in an open to somewhat dense terminal cluster; petals about ⅛ inch (2 to 4 mm) long.

Fruits: Deeply two-lobed, many-seeded capsule.

Ecology: Wet areas, usually along streams, from mid-elevations to the alpine zone, growing more or less throughout the region.

Other Pacific Northwest species have leaves resembling those of brook saxifrage. One such species, **nodding saxifrage** *(S. cernua),* usually has several unbranched stems and basal, round to fan-shaped, deeply toothed leaves. But a distinguishing characteristic of this species is that all but one or a few of the flowers—the terminal one(s)—are replaced by small, red bulbs (bulblets), which ultimately fall to the ground and produce new plants. The few normal flowers are showy, with white, purplish veined petals about ⅜ inch (10 mm) long. The flowers nod before they open, which explains the common name.

Rusty Saxifrage *Saxifraga ferruginea*

Brook Saxifrage
Saxifraga nelsoniana

Nodding Saxifrage
Saxifraga cernua

Yellow Saxifrage *Saxifraga hirculus*

Named for its yellow flowers, this saxifrage has hairy, tufted stems 4 to 8 inches (10 to 20 cm) high.

Leaves: Alternate and basal, linear to club shaped, ⅜ to 1½ inches (1 to 4 cm) long.

Flowers: Bright yellow, often two-toned, solitary on erect stems; petals about ⅜ inch (1 cm) long; sepals eventually hang downward.

Fruits: Two-lobed, many-seeded capsules.

Ecology: A common plant in Alaska and northern British Columbia, where it grows in bogs and moist meadows from lowlands to mid-elevations in the mountains.

Another yellow-flowered saxifrage with a similar range is *S. flagellaris;* it produces runners in the same manner as strawberries.

Purple Mountain Saxifrage *Saxifraga oppositifolia*

This colorful saxifrage grows in very dense, tight cushions, no more than about 1 inch (2 to 3 cm) high.

Leaves: Opposite and overlapping, egg shaped, ⅛ to ¼ inch (3 to 5 mm) long, with short, stiff hairs along the margins.

Flowers: Reddish purple, showy, solitary on short stems; petals about ¼ inch (7 to 9 mm) long; sepals about ⅛ inch (2 to 4 mm) long, with stiff hairs along the margin.

Fruits: Two-lobed, many-seeded capsules.

Ecology: A wide-ranging Arctic species that extends southward along the various mountain ranges, reaching into the North Cascades of Washington; grows in rock crevices and on cliff faces in the alpine zone.

Alpine Saxifrage *Saxifraga tolmiei*

Alpine saxifrage is a low, mat-forming, succulent herb. The flowering stems are erect, 1 to 3 inches (3 to 8 cm) high.

Leaves: Basal and alternate on nonflowering, spreading stems, thick and fleshy, round and lance shaped or somewhat flattened and spoon shaped; ⅛ to ⅜ inch (3 to 10 mm) long.

Flowers: White, mainly solitary on short leafless stems; petals about ¼ inch (4 to 6 mm) long; sepals also white, smaller than the petals.

Fruits: Two- or three-lobed, many-seeded capsules.

Ecology: Often a dominant species on rocky slopes below snowfields in alpine and subalpine zones; distributed throughout the region.

Yellow Saxifrage *Saxifraga hirculus*

Purple Mountain Saxifrage *Saxifraga oppositifolia*

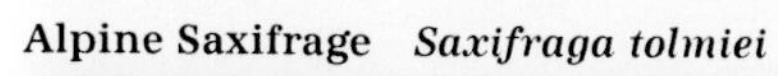

Alpine Saxifrage *Saxifraga tolmiei*

Fringecup

Tellima grandiflora

A common forest species, fringecup has conspicuously hairy leaves and petioles. It grows from rhizomes. The stems are few to several, up to 3 feet (1 m) tall.

Leaves: Mainly basal, becoming progressively smaller upward on the stem; blade heart shaped to triangular, 1 to 4 inches (3 to 10 cm) long and broad, irregularly lobed and toothed; petioles long and densely hairy.

Flowers: Attractive and aptly described by the common name–the white to reddish petals are divided at the tip (fringed) and borne on a green, cup-shaped hypanthium.

Fruits: Two-lobed, many-seeded capsule.

Ecology: Moist lowland forests to subalpine meadows of southeastern Alaska and the western slopes of the Coast Mountains of British Columbia and the Cascades and Olympics in Washington.

Other forest "saxifrages" resemble fringecup (and foamflower) in general growth form. For example, the leaves of **youth-on-age** *(Tolmiea menziesii)* are similar to those of fringecup, but differ in having leaf buds that are produced at the base of the leaf blade ("youth on age"). However, youth-on-age is clearly distinct when in bloom; the hypanthium and sepals are greenish purple, and the petals are chocolate colored and linear. It is similar to fringecup in ecology and distribution, too. Some species of *Heuchera* (alumroot) resemble fringecup, but they have extensively branched flower clusters and small, white flowers. They inhabit rocky sites in moist forests.

Foamflower

Tiarella trifoliata

Foamflower grows from short rhizomes and has stems that are solitary or more often clustered, erect to spreading, and 6 to 15 inches (15 to 40 cm) long.

Leaves: Mainly basal, becoming progressively smaller upward on the stem, simple and three lobed (var. *unifoliata*), trifoliate (var. *trifoliata*), or divided into several segments (var. *laciniata*).

Flowers: White, small, and numerous in narrow racemes; petals linear, about ⅛ inch (2 to 5 mm) long.

Fruits: Few-seeded capsules with two unequal compartments.

Ecology: A conspicuous, often dominant member of forest communities from lowlands to high elevations in the montane zone; in particularly dense populations, the flowers look like foam in the forest understory; distributed in some varietal form more or less throughout the range, especially on the moist, western side of the mountains.

Fringecup *Tellima grandiflora*

Youth-on-Age *Tolmiea menziesii*

Foamflower *Tiarella trifoliata*

Stonecrop Family

Crassulaceae

Many species in this diverse family of succulent plants resemble cacti. Most species grow in dry, subtropical regions of the world, especially in the Old World, and many species are cultivated as house and garden plants, including the well-known poinsettia. The Pacific Northwest has only one genus in the family–*Sedum,* the stonecrops.

The flowers of stonecrops resemble those of saxifrages, differing in that they lack a hypanthium (the floral parts are borne on the receptacle) and they have five rather than two or three pistils. Flowers of both families typically have parts in fives: five petals, five sepals, and ten stamens.

Lance-Leaved Stonecrop

Sedum lanceolatum

This succulent perennial herb spreads by rhizomes and often forms clumps or small mats. The flowering stems are 2 to 8 inches (5 to 20 cm) high.

Leaves: Alternate and basal, linear to lance shaped, round in cross section, about ⅜ inch (1 cm) long.

Flowers: Bright yellow and showy, borne in rather dense clusters at the stem tip; petals pointed, about ¼ inch (6 to 8 mm) long.

Fruits: Numerous-seeded pods, five per flower.

Ecology: A species of open rocky places, from lowlands upward into the alpine zone; distributed throughout the region, common mainly in Washington and southern British Columbia.

Rose Root

Sedum integrifolium (roseum)

The name *rose root* refers to this plant's thick, reddish rhizomes. Its stems are clumped and 2 to 8 inches (5 to 20 cm) high.

Leaves: Alternate, crowded on the stem, fleshy but flattened, narrowly egg shaped, about ½ inch (1 to 2 cm) long, sometimes toothed.

Flowers: Deep reddish purple, densely clustered at the stem tip; unisexual (male and female flowers borne on separate plants); petals only about ⅛ inch (2 to 4 mm) long.

Fruits: Numerous-seeded red pods, five per flower.

Ecology: A common plant of the Arctic tundra in Alaska, extending southward along mountain ranges into Washington and British Columbia; found in rocky, usually moist sites, at least in spring.

Lance-Leaved Stonecrop *Sedum lanceolatum*

Rose Root flowers *Sedum integrifolium*

Rose Root fruits *Sedum integrifolium*

Spreading Stonecrop *Sedum divergens*

The nonflowering stems on this plant spread and root freely, forming mats. The erect, flowering stems are 2 to 6 inches (5 to 15 cm) high.

Leaves: Opposite and basal, broadly egg shaped, flattened but thick and fleshy, ⅛ to ⅜ inch (4 to 10 mm) long and nearly as wide.

Flowers: Bright yellow, in open or dense clusters at the stem tip; petals ¼ to ⅜ inch (6 to 10 mm) long, sharp pointed.

Fruits: Five numerous-seeded pods that spread like the points of a star.

Ecology: An occasional plant in the mountains of Washington and southern British Columbia; grows on rocky ridges and talus slopes in the alpine zone.

In the foothills on the western side of the Washington Cascades and Olympics and the Coast Mountains of southern British Columbia, **broad-leaved stonecrop** *(S. spathulifolium)* is a striking inhabitant of rock outcrops and cliffs. It resembles spreading stonecrop, but the leaves are alternate, larger, and vary in color from gray-green to deep red.

Worm-Leaf Stonecrop *Sedum stenopetalum*

The nonflowering stems on this stonecrop are prostrate and root freely, forming mats. Erect flowering stems are clustered and stand 4 to 8 inches (10 to 20 cm) high. The plant spreads by rhizomes.

Leaves: Alternate and basal, linear to narrowly lance shaped, about ⅛ to ¾ inch (5 to 20 mm) long, becoming twisted and wormlike as they age.

Flowers: Bright yellow, densely clustered at stem tips; petals spreading, sharp-pointed, ¼ to ⅜ inch (6 to 10 mm) long.

Fruits: Numerous-seeded pods, five per flower.

Ecology: Rocky ridges and open forests, mainly on the eastern side of the Washington Cascades and the mountains of southern British Columbia.

Stonecrops are equally at home in the sagebrush steppe and high mountains. High winds, temperature extremes, and periodic drought prevail in both regions. The stonecrops are well adapted to deal with these adverse conditions. They are low-growing, often mat-forming succulents that capably store water for "nonrainy days." They also have an unusual photosynthetic process that allows them to conserve water by keeping stomates closed during the day. They absorb essential carbon dioxide through stomates at night and store it until sunlight energy is available to convert it to sugar.

Spreading Stonecrop *Sedum divergens*

Broad-Leaved Stonecrop *Sedum spathulifolium*

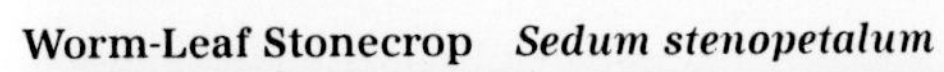

Worm-Leaf Stonecrop *Sedum stenopetalum*

SUNFLOWER FAMILY Asteraceae (Compositae)

Many of the plants familiar to most people are members of the sunflower family—asters, daises, dahlias, thistles, sunflowers, dandelions, zinnias, and goldenrods. Worldwide, it is probably the largest family of flowering plants, with more than 20,000 species. Representatives grow in most habitats on all continents except Antarctica. There are more than 400 species in the Pacific Northwest.

Headlike flower clusters (inflorescences) easily distinguish members of the sunflower family. The head is often mistaken for a single, large flower but is in fact made up of numerous individual flowers on the broadened top of the stem, called the receptacle. Only a few plants in a few other families have this characteristic. The flowers are of two types: ray flowers and disk flowers. The petals of the ray flowers are fused into a single straplike structure that resembles a single petal. The petals of the disk flowers are fused into a tube, with five lobes, or teeth. The heads consist of either ray or disk flowers or a combination of both, in which the ray flowers are always on the outside.

Additional flower parts include the achenes (single-seeded fruits) and the pappus. The pappus, which is sometimes absent, consists of hairs, bristles, or scales and is attached to the top of the achenes. In some cases, the pappus assists in wind dispersal of the seed, as with the dandelion, which has a well-known downy pappus (parachute). The involucre is attached around the rim of the head, and consists of scalelike or somewhat leaflike bracts called involucral bracts. These bracts are sometimes hooked, as in burdock *(Arctium),* and assist in seed dispersal by becoming attached to animals (or clothing).

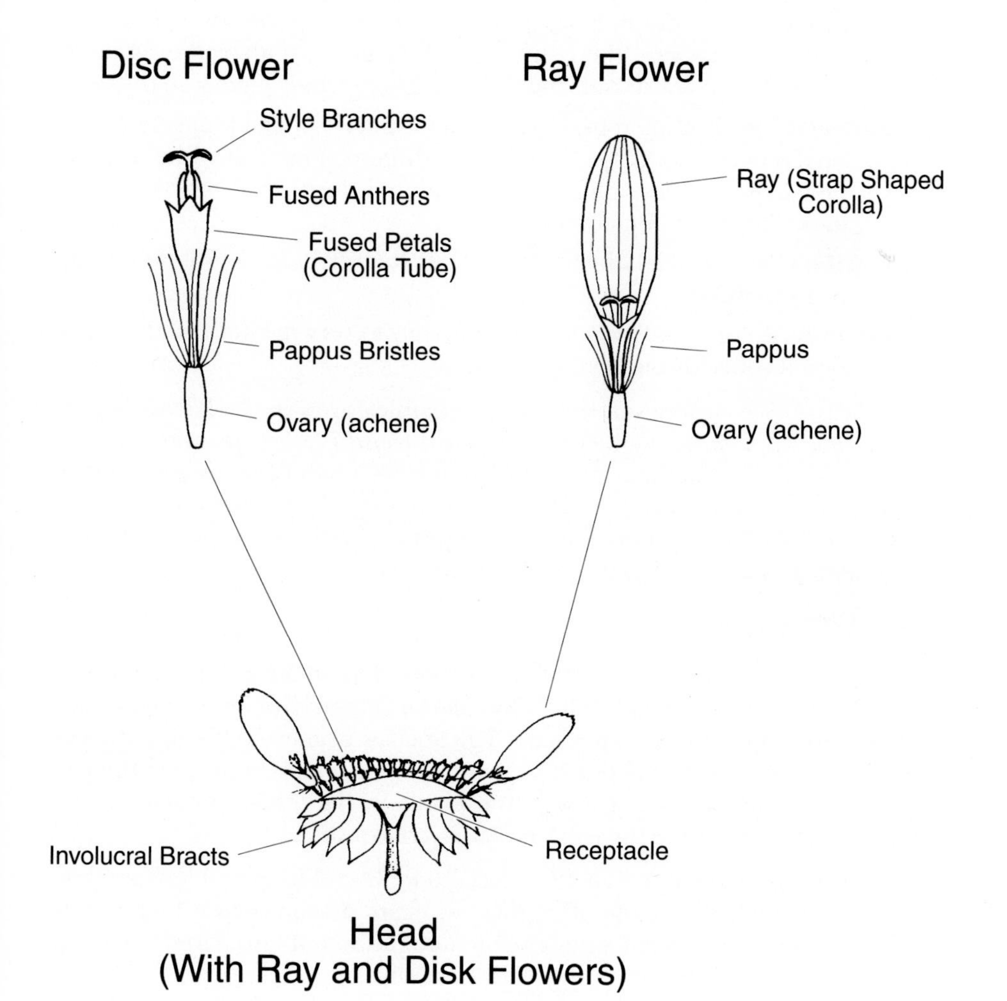

Head
(With Ray and Disk Flowers)

Yarrow

Achillea millefolium

The generic name, *Achillea,* relates to the legend that Achilles first revealed the healing power of this herb and used it to treat soldiers wounded in the siege of Troy. Yarrow is an aromatic perennial herb, usually with well-developed rhizomes. The stems are up to 3 feet (1 m) tall.

Leaves: Basal leaves 1 to 8 inches (2 to 20 cm) long, stalked, pinnately dissected, the divisions again dissected, making the leaves fernlike—*millefolium* means "thousand leaves"; stem leaves alternate, similarly dissected but not stalked and reduced progressively higher on the stem.

Flowers: Heads numerous, borne in hemispheric or flat-topped clusters; ray flowers usually five, sometimes three or four and white to pink; disk flowers ten to thirty, cream colored; involucral bracts light to dark margined, sometimes hairy.

Fruits: Glabrous, flattened achenes, less than ⅛ inch (1 to 2 mm) long; pappus absent.

Ecology: A common species on moderately to very dry sites in all vegetational zones throughout the region.

At least three varieties of this species complex grow in the Pacific Northwest: *alpicola,* a short, high-elevation form; *borealis,* a low to mid-montane form that grows on upland slopes in the northern part of the region, distinguished by green, lightly hairy leaves; and *lanulosa,* a montane form covered with gray hairs that grows on upland slopes in the southern part of the region. Intergradation takes place among all forms.

Pathfinder

Adenocaulon bicolor

This is a perennial herb with fibrous roots. The stems are slender, erect, solitary, and up to a meter tall. White "wool" covers the lower part of the stem, stalked glands the upper part. This species is unusual because it lacks both ray flowers and a pappus. The name *pathfinder* may come from the fact that after someone walks through the plants in the forest understory, the white undersides of the turned-over leaves are conspicuous.

Leaves: Mainly basal, 4 to 12 inches (10 to 30 cm) long and half as wide, long stalked, broadly triangular to heart shaped, green and mainly hairless above, white-woolly below; margins smooth to coarsely toothed; stem leaves alternate, uppermost scalelike.

Flowers: Heads small, few to numerous in a branched, glandular inflorescence; ray flowers absent; disk flowers whitish, only the outer three to seven fertile; involucre about ⅛ inch (2 to 3.5 mm) high, the bracts green, glabrous, equal in size.

Fruits: Club-shaped achenes, becoming about ¼ inch (5 to 8 mm) long, with stalked glands on upper portion; pappus absent.

Ecology: Frequent in moist, shady forests at low to moderate elevations. It ranges from southern British Columbia through Washington.

Yarrow *Achillea millefolium* var. *alpicola*

Pathfinder *Adenocaulon bicolor*

Yarrow *Achillea millefolium* var. *lanulosa*

Orange Agoseris *Agoseris aurantiaca*

This taprooted, perennial herb has milky juice. The stems are unbranched, solitary or few, and grow 4 to 24 inches (10 to 60 cm) tall.

Leaves: All basal, 3 to 14 inches (5 to 35 cm) long, and ⅜ to 1¼ inches (1 to 3 cm) wide, long stalked, linear to lance shaped, hairless to slightly hairy, the margin smooth to sometimes toothed or lobed.

Flowers: Heads solitary; ray flowers burnt orange or rarely yellow, often becoming pink or purple with age; disk flowers absent; involucre ⅝ to ⅞ inch (15 to 25 mm) high, the bracts finely hairy, marginally fringed, and often purple spotted.

Fruits: Achenes, abruptly narrowed to a slender beak, which is one-half to fully as long as the body, about ¼ inch (5 to 9 mm) long overall; pappus of numerous white bristles.

Ecology: Moist to dry meadows and forest openings from upland slopes to the alpine zone. A frequent and widespread species throughout the region.

Short-Beaked Agoseris *Agoseris glauca* var. *dasycephala*

This is another taprooted, perennial herb with milky juice. Slightly taller than orange agoseris, the stems are unbranched, solitary or few, and 4 to 28 inches (10 to 70 cm) tall.

Leaves: All basal, 2 to 12 inches (5 to 30 cm) long, stalked, linear to lance shaped, hairless to sparsely hairy, the margin smooth to wavy, toothed, or lobed.

Flowers: Heads solitary; ray flowers yellow, often becoming pink with age; disk flowers absent; involucre ⅜ to 1¼ inches (1 to 3 cm) high, the bracts slightly to densely hairy, fringed on the margin and sometimes purple spotted.

Fruits: Smooth achenes, tapering to a stout beak up to half as long as the body, sometimes beakless or nearly so; ⅛ to ⅜ inch (5 to 10 mm) long overall; pappus of white bristles.

Ecology: The most common agoseris, growing throughout the region in forest openings (mainly on the eastern side of the mountains) and in moderately to very dry meadows in the subalpine and alpine zones.

A related species with somewhat larger heads is **large-flowered agoseris** *(A. grandiflora).* Its major distinguishing characteristic, however, is the achene, which has a beak two to four times as long as the body of the achene. It is a showy plant, growing in moderately moist to dry, open forests and meadows from lowlands to the subalpine zone. It ranges from the southern Coast Mountains of British Columbia southward through the Cascade Mountains of Washington.

Orange Agoseris
Agoseris aurantiaca

Large-Flowered Agoseris
Agoseris grandiflora

Short-Beaked Agoseris *Agoseris glauca* —Sylvia M. Douglas photo

Pearly Everlasting *Anaphalis margaritacea*

The common name derives from the pearly white involucral bracts, which retain their color and shape when dried and make attractive, long-lasting dried bouquets. A weedy perennial herb up to 3 feet (1 m) tall, this plant spreads by rhizomes. The stems are woolly, white, and usually unbranched.

Leaves: Alternate, 1 to 6 inches (2 to 15 cm) long, narrowly lance shaped, with a conspicuous midvein, greenish above, white-woolly beneath; margins often rolled under.

Flowers: Small heads all of yellowish disk flowers, in dense flat-topped clusters; involucres about ¼ inch (5 to 7 mm) high, the bracts pearly white, with a dark triangular base.

Fruits: Very small, rough, hairless to sparsely hairy achenes.

Ecology: A weedy species of open forests, dry rocky slopes, logged areas, and roadsides; common and widespread throughout the region, from lowlands up to the subalpine zone.

Pearly everlasting is related to, and sometimes confused with, pussytoes (*Antennaria* ssp.) and cudweeds (*Gnaphalium* ssp.). However, pussytoes have shorter and less-developed stem leaves, and cudweeds lack well-developed rhizomes.

Woolly Pussytoes *Antennaria lanata*

This perennial herb, clothed with white "wool," has a branched root crown that gives rise to several erect stems, up to 1 foot (5 to 30 cm) high.

Leaves: Basal leaves 1 to 4 inches (2 to 10 cm) long, lance shaped, densely white- or gray-woolly on both sides; stem leaves are narrower, alternate, and progressively smaller higher on the stem.

Flowers: Heads several in a compact cluster; involucres about ¼ inch (5 to 8 mm) high, the bracts greenish and hairy at the base, brown or greenish black higher up, often white at the tip.

Fruits: Glabrous achenes; pappus hairs white.

Ecology: Common in the southern Coast Mountains of British Columbia and the Cascades of Washington, mainly in areas of late snowmelt, in moderately moist meadows, and on ridges in subalpine and alpine zones.

Pearly Everlasting
Anaphalis margaritacea

Pearly Everlasting
Anaphalis margaritacea

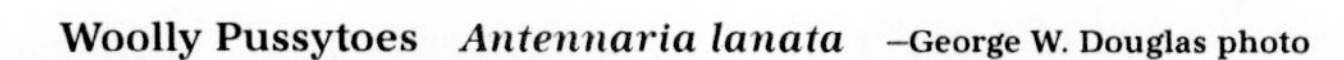

Woolly Pussytoes *Antennaria lanata* –George W. Douglas photo

Alpine Pussytoes *Antennaria alpina* var. *media*

Antennaria species are called "pussytoes" because the densely clustered flowering heads, especially when in fruit, look like furry cat paws. This perennial herb spreads by short stolons, forming dense mats. The stems are up to 6 inches (1 to 14 cm) high.

Leaves: Basal leaves up to 1½ inches (0.3 to 4 cm) long, broadly lance shaped, or spoon shaped, and densely covered with white or gray woolly hairs on both sides; stem leaves alternate, linear, and smaller higher on the stem.

Flowers: Heads several (two to nine) in a tight, rounded inflorescence; ray flowers absent; disk flowers whitish; involucres ⅛ to ⅜ inch (3 to 10 mm) high, the bracts woolly below and dark brownish green to black above; flowers unisexual (male and female flowers growing on separate plants).

Fruits: Hairless achenes, sometimes with small bumps; pappus of white bristles.

Ecology: Distributed throughout the Pacific Northwest in a variety of habitats, from moist alpine snow beds to dry subalpine and alpine ridges.

Distinguishing alpine pussytoes from **rosy pussytoes** *(A. microphylla)* and **umber pussytoes** *(A. umbrinella)* is often difficult. The three species are separated mainly by the color of the involucral bracts. Alpine pussytoes has brownish green to black bracts. The bracts are white to pink or rosy red in rosy pussytoes and brownish in umber pussytoes. Also, rosy pussytoes is typically somewhat larger than alpine pussytoes in all respects and umber pussytoes is a bit smaller than alpine pussytoes.

Alpine and umber pussytoes are common species, particularly in the southern part of the region. Both species grow in a variety of habitats, including moderately to very dry open forests and subalpine and alpine meadows and ridges.

In the northern mountains, alpine pussytoes may grow together with **one-headed pussytoes** *(A. monocephala)*, a species with similar colored involucral bracts. It is distinguished by its solitary (or rarely two) flower heads.

The taxonomy of the genus *Antennaria* is further complicated because species often produce seeds without fertilization (by agamospermy)–seeds without sex. The result is a complex mix of sexual and asexual plants.

Alpine Pussytoes *Antennaria alpina*

Rosy Pussytoes *Antennaria microphylla*
—George W. Douglas photo

Umber Pussytoes *Antennaria umbrinella*

Heart-Leaved Arnica *Arnica cordifolia*

Sparsely hairy to often glandular, this perennial herb spreads by long rhizomes. The stems are erect, solitary or sometimes clustered, and stand 4 to 24 inches (10 to 60 cm) tall.

Leaves: Basal and opposite, the latter in two or three pairs; the larger, lower pair of opposite leaves has long petioles and heart-shaped blades 1 to 5 inches (2 to 12 cm) long; upper leaves lance shaped and smaller; leaf margins smooth to coarsely toothed.

Flowers: Heads one to three or sometimes more; ray flowers yellow, usually with apical teeth; disk flowers also yellow; involucres ½ to ⅞ inch (13 to 22 mm) high; involucral bracts lance shaped to elliptical, pointed, sparsely to densely covered with long, white hair and often sticky.

Fruits: Achenes covered with stiff hairs, sometimes glandular; pappus of white, barbed hairs.

Ecology: Widespread throughout the region; grows in almost every forest type and less frequently extends upward into the subalpine and alpine zones, particularly northward.

This species occasionally hybridizes with mountain arnica, resulting in intermediate plants.

Mountain Arnica *Arnica latifolia*

Also called broad-leaf arnica, this perennial, rhizomatous herb has stems that are erect, solitary or clustered, sparsely hairy, and 4 to 24 inches (10 to 60 cm) tall.

Leaves: Basal and opposite, the latter in two to five pairs; leaf blades 1 to 6 inches (2 to 15 cm) long, ⅛ to 3 inches (0.5 to 8 cm) wide, lance shaped to elliptical, coarsely toothed, sparsely to moderately hairy, and usually glandular; lower leaves stalked; upper leaves smaller and nonstalked.

Flowers: Heads one to five; the ray and disk flowers yellow; involucres ¼ to ⅝ inch (7 to 18 mm) high, the bracts lance shaped to elliptical, sparsely to densely hairy, often glandular, with fringed tips.

Fruits: Hairless to hairy achenes; pappus of white, barbed hairs.

Ecology: Inhabits moist forests, streamsides, and subalpine and alpine meadows; common in the mountains of Washington and southern British Columbia, infrequent farther north.

Heart-Leaved Arnica *Arnica cordifolia*

Mountain Arnica *Arnica latifolia*

Mountain Arnica *Arnica latifolia*

Northern Arnica *Arnica frigida*

A hairy to glandular-hairy perennial herb, its stems are erect, usually solitary, and 2 to 16 inches (5 to 40 cm) tall.

Leaves: Basal and opposite, the latter in one to four pairs, ½ to 4 inches (1.5 to 10 cm) long, lance shaped, hairless to long-hairy; the basal and lower stem leaves stalked and toothed.

Flowers: Heads usually solitary, rarely three; ray flowers yellow and toothed at the tip; disk flowers yellow; involucres ⅜ to ¾ inch (10 to 19 mm) high, the bracts lance shaped and sparsely to densely covered with long hairs below, becoming hairless above.

Fruits: Hairless to sparsely hairy achenes; pappus of white, barbed hairs.

Ecology: A plant of the extreme north, mainly on high, dry, alpine slopes and ridges or at lower elevations on moist, northerly slopes.

Another species that grows in the extreme northern part of the region, in the northern Coast Mountains of British Columbia, is **purple arnica** *(A. lessingii).* It has solitary stems and three to five pairs of lance-shaped leaves. The flower heads are solitary and nodding. The common name describes the purple anthers, which, together with the nodding head, characterize this attractive species. It grows in moist sites, such as late snowmelt areas.

Parry's Arnica *Arnica parryi*

Because the heads lack ray flowers, this arnica is easy to identify. It is a perennial, hairy to glandular-hairy, rhizomatous herb. The stems are erect, solitary, sometimes branched, and stand 6 to 24 inches (15 to 60 cm) tall.

Leaves: Basal and opposite, in two to four pairs, 1 to 10 inches (3 to 25 cm) long, lance shaped, stalked, and sometimes toothed.

Flowers: Heads usually one to four, sometimes up to twelve; ray flowers usually absent; disk flowers yellow; involucres ⅗ to ⅝ inch (10 to 14 mm) high, with bracts lance shaped, hairy, and glandular.

Fruits: Hairless to hairy achenes; pappus of tawny to brownish, barbed to feathery hairs.

Ecology: Grows in montane forests and subalpine meadows; common in the mountains of Washington and southern British Columbia; rare farther north.

Northern Arnica *Arnica frigida*
—George W. Douglas photo

Parry's Arnica *Arnica parryi*

Purple Arnica *Arnica lessingii*

Mountain Sagewort

Artemisia norvegica ssp. *saxatilis (A. arctica)*

This densely hairy to nonhairy perennial herb has one to several erect stems, 8 to 24 inches (20 to 60 cm) tall. Frequent short runners bear stemless rosettes of leaves. Unlike most species of *Artemisia,* mountain sagewort does not have a sagelike smell.

Leaves: Basal leaves 1 to 4 inches (2.5 to 10 cm) long, not including the petiole, pinnately divided into linear segments, hairless to densely covered with long hair; stem leaves alternate, few, unstalked, and progressively reduced higher up the stem.

Flowers: Heads numerous, up to ⅜ (10 mm) wide, often nodding in spikelike clusters; ray flowers absent; disk flowers yellow to reddish; involucres ⅛ to ½ inch (3 to 7 mm) high, with bracts hairless or moderately covered with long hairs, greenish with dark margins.

Fruits: Hairless achenes; pappus absent.

Ecology: Common on rocky slopes and meadows in the subalpine and alpine zones throughout the Pacific Northwest; occasionally grows in northern forest openings.

Leafy Aster

Aster foliaceus

This common perennial herb is rhizomatous and varies considerably in height, growth form, and hairiness. It grows from 4 to 24 inches (10 to 60 cm) tall.

Leaves: Basal and alternate, lance shaped to rounded, 1 to 5 inches (2.5 to 12 cm) long, smooth margined; basal leaves usually have stalks; stem leaves clasp the stem and have earlike lobes at the leaf base.

Flowers: Heads one to many; ray flowers fifteen to sixty, rose-purple to blue or violet; disk flowers yellow; involucres about ⅜ inch (9 to 12 mm) high, the bracts overlap, some of them leafy.

Fruits: Hairless to fine hairy achenes; pappus of white to brownish bristles.

Ecology: Grows throughout the region in moist to dry sites in all vegetational zones.

This variable species is difficult to identify because it is similar to several other asters. Smooth aster *(A. laevis)* has smaller flower heads, Douglas' aster *(A. subspicatus)* has yellowish or brownish involucral bracts, and western aster *(A. occidentalis)* has much narrower leaves.

Alpine aster *(A. alpigenus)* resembles some forms of leafy aster. An attractive plant, its large solitary flower heads have lavender rays and yellow disk flowers. Several spreading stems arise from its cluster of grasslike leaves. It grows in the alpine and subalpine zones of the Cascade and Olympic Mountains of Washington.

Mountain Sagewort
Artemisia norvegica

Leafy Aster
Aster foliaceus

Alpine Aster *Aster alpigenus*

Cascade Aster *Aster ledophyllus*

This leafy, perennial herb, usually with several erect stems, grows up to 3 feet (1 m) tall, from a stout, branched root crown.

Leaves: Mainly alternate, 1 to 3 inches (2 to 7 cm) long, lance shaped to broadly elliptical, unstalked, usually smooth margined, hairless to slightly hairy above, gray-cottony below.

Flowers: Heads usually several; ray flowers thirteen to twenty-one, lavender-purple; disk flowers yellow; involucres about ⅜ inch (7 to 11 mm) high, the bracts linear to lance shaped, pointed, and overlapping, somewhat greenish above, and often purplish at the tip.

Fruits: Stiff-hairy achenes; pappus of white bristles.

Ecology: Frequent in forest openings and subalpine and alpine meadows; in the Pacific Northwest, it grows only in the Washington Cascades.

When flowers are lacking, this species could be mistaken for **Engelmann's aster** *(A. engelmannii),* a taller plant with broader leaves. The ray flowers, usually thirteen, are white, often becoming pink with age, and the disk flowers are yellow. Engelmann's aster is a frequent species in open forests and subalpine meadows in the Cascade Mountains of Washington and adjacent British Columbia.

Arctic Aster *Aster sibiricus* ssp. *meritus*

A high-elevation perennial herb, arctic aster spreads by slender, creeping rhizomes. The stems are 2 to 16 inches (5 to 40 cm) tall, solitary or few, sometimes branched, and more or less hairy.

Leaves: Basal leaves small, firm, and stalked; stem leaves alternate, mainly stalked, at least on the flower stem, ½ to 4 inches (1.5 to 10 cm) long, lance shaped to elliptical, progressively reduced upward; margins smooth or with a few small teeth; short hairs cover the undersurface.

Flowers: Heads one to twenty; ray flowers ten to twenty-five, purple; disk flowers yellow; involucres ¼ to ⅝ inch (7 to 15 mm) high, the bracts oblong to lance shaped and somewhat hairy—the outer ones green and the inner ones purplish or purple tipped.

Fruits: Hairy achenes; pappus of brown or yellowish hairs.

Ecology: Grows in moderately moist to dry alpine meadows and on scree slopes and rocky ridges from Alaska southward to northern Washington.

Cascade Aster *Aster ledophyllus*

Englemann's Aster *Aster engelmannii*

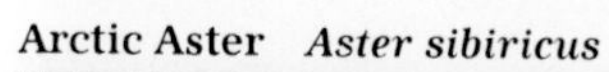

Arctic Aster *Aster sibiricus*

Edible Thistle *Cirsium edule*

This spiny, robust biennial or short-lived perennial herb grows from a taproot. The thick, hairy, and rather succulent stems stand from 1 to 6 feet (up to 2 m) tall.

Leaves: Basal and alternate, up to 16 inches (40 cm) long, lance shaped but irregularly lobed and toothed, spiny, green, and sparsely hairy on both surfaces.

Flowers: Heads solitary to many, borne at the ends of short branches; ray flowers absent; disk flowers pinkish purple with long tubular bases; involucres 1 to 2 inches (2.5 to 5 cm) high and covered with tangled cottony hairs; the outer involucral bracts have short spines.

Fruits: Hairless, prismatic, several-veined achenes; pappus of feathery white bristles.

Ecology: Common in the southern half of the region, less so to the north in southeastern Alaska and absent from the Queen Charlotte Islands; found in moist to moderately dry forest openings and subalpine meadows.

Short-styled thistle *(C. brevistylum)* resembles edible thistle, but its styles are shorter (at most, 1/16 inch [1.5 mm] longer than the disk flowers as opposed to 1/8 inch [3 mm] in edible thistle), and the leaves are lobed less than one-half the width of the leaf blade.

Bitter Daisy *Erigeron acris*

Bitter daisy, a hairy, perennial herb, grows from fibrous roots. The stems are erect, solitary to several, and up to 30 inches (20 to 80 cm) tall.

Leaves: Basal and alternate, up to 6 inches (15 cm) long; basal leaves broadly lance shaped to spoon shaped, stalked, and smooth margined; stem leaves narrower and shorter.

Flowers: Heads few to numerous in a flat to round-topped cluster; ray flowers numerous, pink to purplish or white and very small, about 1/8 inch (2.5 to 4.5 mm) long; disk flowers yellow; involucres about 1/4 inch (5 to 12 mm) high with bracts nearly equal in length, and covered with hairs.

Fruits: Sparsely hairy, two-veined achenes; pappus of slender, barbed white or more or less reddish bristles.

Ecology: Distributed throughout the region on moist to dry sites in all vegetational zones.

This species could be mistaken for **spear-leaved daisy** (*E. lonchophyllus*), but unlike bitter daisy, its inflorescence is elongate.

Edible Thistle *Cirsium edule*

Edible Thistle *Cirsium edule*

Bitter Daisy *Erigeron acris*

Cut-Leaved Daisy *Erigeron compositus* var. *glabratus*

Widespread and variable, this glandular-hairy, perennial herb has one to several erect, 1 to 10 inch (2.5 to 25 cm) stems derived from a branched root crown.

Leaves: Basal, up to 3 inches (0.5 to 8 cm) long, hairy, divided at the tip, and fernlike; stem leaves few and bractlike.

Flowers: Heads solitary; ray flowers twenty to sixty, white to pale lavender, ⅛ to ½ inch (4 to 12 mm) long; disk flowers yellow; involucres about ¼ inch (5 to 10 mm) high, the bracts lance shaped, glandular, and covered with spreading hairs.

Fruits: Hairy achenes; pappus of white bristles.

Ecology: A frequent species throughout the region; grows in a variety of well-drained sites, from gravelly riverbanks in the montane zone to dry, exposed subalpine and alpine meadows, and gravelly ridgetops.

Salish daisy *(E. salishii)* closely resembles cut-leaved daisy, especially var. *discoideus,* which has three-pronged leaves with narrow leaflets. However, the leaves of Salish daisy persist for several years, are only ⅛ to ⅝ inch (3 to 16 mm) long, and have three rather broad, equal-length prongs. Salish daisy grows on high alpine scree slopes and ridges. It is known only from the central Vancouver Island Ranges and the North Cascades of Washington.

Subalpine Daisy *Erigeron peregrinus*

This common perennial herb has a branched rootstalk. The stems are solitary or few, unbranched, hairy or hairless, and 4 to 24 inches (10 to 60 cm) tall.

Leaves: Basal and alternate, 1 to 8 inches (2.5 to 20 cm) long, narrowly or broadly lance shaped to spoon shaped, usually hairless or hairy margined; upper leaves often clasp the stem.

Flowers: Heads usually solitary; ray flowers pink, lavender, reddish purple, or sometimes whitish, about ¼ to 1 inch (8 to 25 mm) long; disk flowers yellow; involucres about ¼ to ½ inch (6 to 12 mm) high, with bracts lance-oblong, covered with long hairs and/or gland-tipped hairs, which make them sticky.

Fruits: Sparsely hairy achenes; pappus of white to tan hairs.

Ecology: Inhabits wet to moist meadows and open forests; distributed throughout all vegetational zones; especially abundant in the southern half of the region.

This extremely variable species has two recognized subspecies growing in the Pacific Northwest: ssp. *peregrinus,* which has involucral bracts covered with long hairs, and the ssp. *callianthemus,* which has involucral bracts made sticky by glandular hairs.

Cut-Leaved Daisy *Erigeron compositus*

Subalpine Daisy *Erigeron peregrinus*

Salish Daisy *Erigeron salishii* –George W. Douglas photo

Golden Daisy *Erigeron aureus*

A meadow splashed with golden daisy is just one of many alpine wonders, and this plant also makes a splendid subject for the alpine rock garden. It is a hairy perennial herb with a few to several erect to spreading stems, 1 to 6 inches (2.5 to 15 cm) high.

Leaves: Basal and alternate, basal leaves broadly elliptical to rounded, long stalked, the blades about 1 inch (2 to 2.5 cm) long; stem leaves few, narrow, and much smaller.

Flowers: Heads solitary; ray flowers twenty-five to seventy yellow, about ¼ inch (6 to 9 mm) long; disk flowers yellow; involucres about ¼ inch (5 to 8 mm) high, with bracts linear to lance shaped, equal in size, and sparsely to densely covered with dark "wool."

Fruits: Hairy achenes; pappus of white bristles and a few, outer, narrow scales.

Ecology: A common species in dry meadows, on ridges, and in exposed, rocky sites of the upper subalpine and alpine zones; occasionally grows in open, high-elevation forests; ranges through the mountains of Washington into southern British Columbia, primarily on the eastern, drier slopes.

From a short distance, golden daisy could be mistaken for Lyall's goldenweed, another high-elevation yellow-flowered plant. The sticky, glandular leaves of goldenweed and the woolly involucral bracts of golden daisy readily separate these two alpine gems.

Lyall's Goldenweed *Haplopappus lyallii*

A small, common perennial herb that spreads by rhizomes. The stems are erect and 1 to 6 inches (2.5 to 15 cm) high. The entire plant is covered by tiny gland-tipped hairs, which make it sticky.

Leaves: Basal and alternate, ½ to 3 inches (1.5 to 7 cm) long, broadly lance shaped to spoon shaped, not stalked or toothed; upper stem leaves small and bractlike.

Flowers: Heads solitary; ray flowers thirteen to thirty-five, yellow, about ⅜ inch (6 to 11 mm) long; disk flowers yellow; involucres about ⅜ inch (6 to 12 mm) high, with bracts nearly equal in length, purplish, and densely glandular.

Fruits: Mainly hairless achenes; pappus of tawny bristles.

Ecology: Common from the eastern side of the southern Coast Mountains of British Columbia southward through the Cascade Mountains of Washington; found on scree slopes and gravelly ridges, mainly in the alpine zone.

Golden Daisy ***Erigeron aureus***

Lyall's Goldenweed ***Haplopappus lyallii***

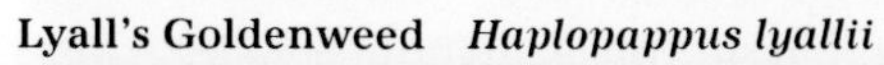

White-Flowered Hawkweed *Hieracium albiflorum*

A hairy, perennial herb with milky juice, white-flowered hawkweed has stems that are solitary to clumped, erect, and 1 to 4 feet (30 cm to 1.2 m) tall.

Leaves: Mainly basal, 2 to 7 inches (4 to 18 cm) long, elliptical to broadly lance shaped, stalked, bristly with short, rigid hair on the upper surface; margins smooth or toothed; stem leaves, if present, alternate and much reduced.

Flowers: Heads several to many in an open cluster; ray flowers white; disk flowers absent; involucres about ¾ inch (6 to 11 mm) high, with bracts linear to lance shaped, nearly equaling each other in length, greenish or blackish, and usually glandular-hairy.

Fruits: Hairless achenes; pappus of white or tannish bristles.

Ecology: Inhabits moist to moderately dry open forests, meadows, clearings, and roadsides in the montane zone, throughout all but the most northern part of the region.

Slender Hawkweed *Hieracium gracile*

This is a rather delicate, hairy or glandular-sticky rhizomatous herb with milky juice. The stems are solitary or few in a cluster, erect, usually unbranched, and 3 to 10 inches (8 to 25 cm) high.

Leaves: Mainly basal, broadly lance shaped to spoon shaped, stalked, ½ to 5 inches (1 to 13 cm) long; margin smooth or toothed; stem leaves absent or few and much reduced.

Flowers: Heads, few to several in a small cluster; ray flowers yellow; disk flowers absent; involucres ⅛ to ⅜ inch (4 to 10 mm) high, with bracts nearly equal in length. Black glandular hairs cover the involucre and upper stem.

Fruits: Hairless achenes; pappus of tannish or white bristles.

Ecology: Inhabits moist snow-bed sites and moderately dry meadows and slopes in subalpine and alpine zones; common in the southern half of the region, less so northward.

Slender hawkweed is easily mistaken for the closely related **woolly hawkweed** *(H. triste).* The latter, however, has larger heads and longer, grayish to grayish black hairs and lacks glands. Woolly hawkweed is common from Alaska south to the central Coast Mountains of British Columbia and the Queen Charlotte Islands, where it often intergrades with slender hawkweed.

White-Flowered Hawkweed *Hieracium albiflorum*

White-Flowered Hawkweed
Hieracium albiflorum

Slender Hawkweed
Hieracium gracile

Oxeye Daisy *Leucanthemum vulgare (Chrysanthemum leucanthemum)*

Attractive and widespread, this perennial herb, usually with several stems 8 to 30 inches (20 to 80 cm) tall, derives from a branched root crown.

Leaves: Basal and alternate, 2 to 6 inches (4 to 15 cm) long, lance shaped to spoon shaped, toothed or cleft; upper stem leaves smaller and nonstalked.

Flowers: Heads solitary at the ends of branches; ray flowers white, ½ to ¾ inch (12 to 20 mm) long; disk flowers yellow; involucres about ⅜ inch (7 to 11 mm) high, with bracts narrowly lance shaped and overlapping.

Fruits: Black achenes; pappus absent.

Ecology: An invasive weed of mountain meadows; usually grows along roadsides and in other disturbed sites; common in the southern part of the region, less so northward.

Sweet Coltsfoot *Petasites frigidus*

Quinault Indians in Washington State once used sweet coltsfoot leaves to cover berries in steam-cooking ground ovens. It is a perennial herb that spreads from rhizomes. The stems are up to 2 feet (60 cm) tall and more or less covered with white hairs.

Leaves: Basal leaves long stalked, triangular to heart shaped, up to 16 inches (40 cm) wide, white-woolly beneath, arising directly from the rhizomes after the plant has flowered; leaf margins variously lobed and toothed; stem leaves alternate and reduced to lance-shaped bracts ⅜ to 2½ inches (1 to 6 cm) long.

Flowers: Heads several to many in a round or flat-topped cluster; more or less unisexual; ray flowers whitish or pink to purplish; disk flowers pink to purplish; involucres ¼ to ½ inch (6 to 12 mm) high, glandular to woolly below, bracts lance shaped to elliptical, the tips tufted with hair.

Fruits: Mainly hairless achenes; pappus of white bristles.

Ecology: Three varieties of this taxonomically complex species grow in the Pacific Northwest; var. *nivalis,* with shallowly cleft leaves, grows in wet to moist meadows, seepage areas, and along stream banks and lakeshores in the subalpine to alpine zones throughout the region; var. *frigidus,* with blunt-toothed leaves, grows in similar habitats but is mainly restricted to northern mountains; var. *palmatus,* with deeply cleft, palmate leaves, grows in moist to wet forests, swamps, meadows, and along stream banks and roadsides at low elevations.

Oxeye Daisy
Leucanthemum vulgare

Sweet Coltsfoot
Petasites frigidus var. *palmatus*

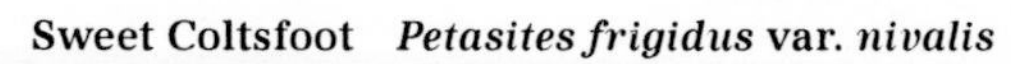

Sweet Coltsfoot *Petasites frigidus* var. *nivalis*

Narrow-Leaf Sawwort *Saussurea angustifolia*

A perennial herb with a branched, woody root crown. The stems are erect and 6 to 18 inches (15 to 45 cm) tall. Patches and lines of tangled white "wool" cover the stems and leaves.

Leaves: Basal and alternate, linear to narrowly elliptical, often partly rolled up, covered with white "wool," especially along the margin; 1 to 4 inches (2 to 10 cm) long.

Flowers: Heads several, clustered at the stem tip; ray flowers absent; disk flowers about thirteen, purplish, about ⅜ inch (10 to 12 mm) long; involucres about ½ inch (9 to 13 mm) high; the bracts in three or four rows, lance shaped, blackish, and tufted with white hair.

Fruits: Hairless achenes; pappus of tannish, feathery hairs.

Ecology: A tundra species that reaches into the Pacific Northwest in southeastern Alaska; inhabits dry sites in the high mountains.

American sawwort *(S. americana)*, a related plant and much more robust, is distributed over much of the region. It has triangular, toothed leaves resembling arrow-leaved groundsel. American sawwort grows mainly in lush, moist, well-drained subalpine meadows or avalanche tracks.

Western Groundsel *Senecio integerrimus*

Western groundsel is a perennial herb with fibrous roots. The stems are mainly solitary, erect, and up to 3 feet (1 m) tall. Tangled, white "wool" often occurs on the stem and leaves.

Leaves: Basal and alternate, basal leaves 2 to 10 inches (5 to 25 cm) long, stalked, somewhat succulent, lance shaped to spoon shaped, margin smooth to irregularly toothed; the stem leaves are progressively reduced upward and clasp the stem.

Flowers: Heads several to numerous, often tightly clustered; ray flowers yellow, ¼ to ⅝ inch (6 to 16 cm) long or sometimes absent; disk flowers yellow; involucres about ¼ inch (5 to 10 mm) high, with bracts lance shaped and usually black tipped, sometimes hairy.

Fruits: Hairless or stiff hairy achenes; pappus of white hairs.

Ecology: Grows in moist to moderately dry open forests and meadows, from lowlands to the subalpine zone along the eastern side of the Washington Cascades and southern Coast Mountains of British Columbia.

Elmer's butterweed *(S. elmeri)* resembles western groundsel but has more prominently toothed leaves and a sprawling habit. It is restricted to moist talus and gravelly slopes in the subalpine and alpine zones of the southern Coast Mountains of British Columbia and the Cascades of northern Washington.

Narrow-Leaf Sawwort
Saussurea angustifolia

Western Groundsel
Senecio integerrimus

Elmer's Butterweed *Senecio elmeri* –George W. Douglas photo

Arrow-Leaved Groundsel *Senecio triangularis*

A mainly hairless, perennial herb with fibrous roots, its stems are erect, 1 to 4 feet (30 cm to 1.2 m) tall, and clustered.

Leaves: Basal and alternate, 1 to 4 inches (2.5 to 10 cm) long and half as wide, triangular, strongly toothed, stalked (except the upper stem leaves), more or less hairless.

Flowers: Heads few to numerous in a short, flat-topped cluster; ray flowers yellow, usually about eight; disk flowers yellow; involucres about ⅜ inch (7 to 12 mm) high, the bracts greenish, with tufts of black hair on the tips.

Fruits: Hairless achenes; pappus of white hairs.

Ecology: Distributed throughout the region, growing in moist to wet meadows, avalanche tracks, open forests, and along stream banks, especially in the subalpine zone.

Northern Groundsel *Senecio tundricola*

A white-woolly, perennial herb, northern groundsel has erect, usually solitary stems 1 to 12 inches (2.5 to 30 cm) high.

Leaves: Basal and alternate, 1 to 4 inches (2 to 10 cm) long, short stalked, broadly lance shaped, smooth margined; upper leaves smaller and nonstalked.

Flowers: Heads usually solitary; ray flowers orange-yellow; disk flowers yellow; involucres about ⅜ inch (6 to 12 mm) high, with bracts linear to lance shaped, covered with purplish white "wool."

Fruits: Stiff, hairy achenes; pappus of white hairs.

Ecology: A frequent species of dry alpine meadows, gravelly slopes, and ridges in northern British Columbia and Alaska.

Several other Se*necio* species grow in the Pacific Northwest. **Rayless alpine butterweed** *(S. pauciflorus)* is a common species in wet subalpine meadows. It has small, clustered heads with no ray flowers and orange to reddish disk flowers; the involucre is reddish purple. It ranges throughout the region except in the Queen Charlotte Islands.

Rayless mountain butterweed *(S. indecorus)* is another rayless species distributed throughout the region. It has green involucres and yellow disk flowers.

Black-tipped groundsel *(S. lugens)* resembles western groundsel but has regularly toothed basal leaves and is mainly an alpine species in the Olympic and northern Cascade Mountains. It shows up sporadically on the eastern side of the Coast Mountains from British Columbia to Alaska.

Arrow-Leaved Groundsel
Senecio triangularis –George W. Douglas photo

Northern Groundsel
Senecio tundricola –Sylvia M. Douglas photo

Rayless Alpine Butterweed *Senecio pauciflorus*

Northern Goldenrod *Solidago multiradiata*

A variable alpine perennial herb with short, branching rhizomes, its stems are usually solitary, erect or spreading, and 2 to 20 inches (5 to 50 cm) tall. At least the upper part of the stem is hairy.

Leaves: Basal and alternate, 1 to 4 inches (2 to 10 cm) long, stalked, broadly lance shaped to spoon shaped, usually toothed, margins conspicuously fringed, at least along the stalks; upper stem leaves smaller and nonstalked.

Flowers: Heads few to many, in loose to dense, short-branched rounded clusters; ray flowers yellow, about thirteen; disk flowers yellow,; involucres about ¼ inch (4 to 6 mm) high; the bracts linear to lance shaped, pointed, not conspicuously overlapping, and more or less fringed.

Fruits: Achenes with short hairs; pappus of white bristles.

Ecology: Distributed over a wide variety of habitats–from moist to dry and from gravelly to rocky–in all vegetational zones; a common species throughout the region, especially in the alpine zone.

Northern goldenrod is sometimes difficult to tell from **spikelike goldenrod** *(S. spathulata),* which differs by having an elongate inflorescence and does not have hairs on the leaf margins.

Spikelike Goldenrod *Solidago spathulata*

A low perennial herb with short, branching rhizomes. The stems are 4 to 24 inches (10 to 60 cm) tall, usually solitary, spreading to erect, and often glutinous toward the top.

Leaves: Basal and alternate, ½ to 6 inches (1.5 to 15 cm) long, broadly lance shaped to spoon shaped or rounded, usually toothed, at least on the upper half, hairless, and stalked; upper stem leaves smaller and nonstalked.

Flowers: Heads numerous, in long, narrow, branched clusters; ray flowers yellow, usually eight; disk flowers yellow; involucres about ¼ inch (4 to 6 mm) high, with bracts overlapping and lance shaped but blunt.

Fruits: Densely hairy achenes; pappus of white bristles.

Ecology: Inhabits open dry forests, gravel riversides, terraces, stream banks and meadows in all vegetational zones; a frequent species throughout the region.

Northern Goldenrod *Solidago multiradiata*

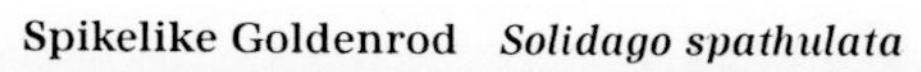

Spikelike Goldenrod *Solidago spathulata*

Sweet Gale Family Myricaceae

Sweet gale is the only Pacific Northwest species of this family.

Sweet Gale *Myrica gale*

Fragrant and deciduous, this is a medium-sized shrub, up to about 6 feet (2 m) tall.

Leaves: Alternate, more or less elliptical but broader toward the toothed tip, 1 to 2 inches (3 to 6 cm) long. Yellowish glands on the leaf surface give the plant a distinctive, sweet odor.

Flowers: Individually inconspicuous, lacking petals and sepals; unisexual (male and female flowers borne on separate plants) and densely clustered with bracts in round to cylindrical catkins that resemble small cones.

Fruits: Somewhat fleshy; one seeded.

Ecology: Inhabits bogs and marshes in the foothills along the western side of the Cascades and Coast Mountain ranges; distributed from Washington to southeastern Alaska, where it is especially common.

Valerian Family Valerianaceae

Valarians comprise a small family of herbaceous plants with opposite leaves. The flowers are small but often showy, in dense clusters. The petals are fused into a narrow tube with five lobes; the stamens, only three, are fused to the petal tube; the sepals are lacking or modified into feathery bristles; and the ovary is borne below the other flower parts (inferior).

Sitka Valerian *Valeriana sitchensis*

A perennial herb with squarish, rather succulent stems and short, thick rhizomes, it grows up to 3 feet (1 m) tall.

Leaves: Opposite and basal, pinnately divided into toothed or lobed leaflets, the terminal leaflet the largest.

Flowers: White to pinkish, up to ⅜ inch (1 cm) long, borne in flat-topped or dome-shaped inflorescences at the stem and branch tips; stamens and style extend beyond the petal tube.

Fruits: One-seeded achenes, parachute-like, with the persistent sepals modified into numerous feathery bristles.

Ecology: A common and conspicuous plant in moist mountain meadows and open forests throughout the region.

Sweet Gale *Myrica gale*

Sitka Valerian *Valeriana sitchensis*

Sitka Valerian *Valeriana sitchensis*

VIOLET FAMILY

Violaceae

Members of this family are easily recognized by their pansylike, bilaterally symmetrical flowers adapted for bumblebee pollination. At the base of the five petals is a small opening into a saclike nectary formed by the modification of the lower petal. Nectar guides on the petals, particularly the lower petal, direct bees to the opening. The flowers have five sepals and five stamens, and the stamens are wrapped around the pistil.

Early Blue Violet

Viola adunca

This low perennial is an early flowering herb with spreading rhizomes. The stems are short, 1 to 4 inches (2 to 10 cm) high.

Leaves: Mainly basal; blades egg shaped to heart shaped, sometimes toothed, 1 to 2 inches (2 to 5 cm) long, with petioles much longer.

Flowers: Blue to dark violet; petals about ⅜ inch (8 to 12 mm) long, the lower one or three with a white base and dark purple lines; the spur is about ¼ inch (6 mm) long.

Fruits: Many-seeded, explosive capsules.

Ecology: A variable species that is adapted to numerous habitats, from dry prairies to rocky alpine ridges; also grows in moist to dry forests at various elevations; distributed throughout the Pacific Northwest.

Marsh Violet

Viola palustris

This perennial herb spreads by both rhizomes and stolons. The flowering stems are leafless and grow 2 to 4 inches (5 to 10 cm) high.

Leaves: Basal, derived from rhizomes or stolons, broadly heart shaped, blades 1 to 2 inches (2 to 5 cm) long and broad.

Flowers: Pale blue-violet; about ½ inch (10 to 15 mm) long, with a prominent nectar spur, solitary on erect, leafless stems; petals typically swept backward.

Fruits: Explosive, many-seeded capsules.

Ecology: A common inhabitant of marshes and bogs from low elevations upward into the subalpine zone in the mountains of Washington and southern British Columbia.

Another blue violet in the region is **Aleutian violet** *(V. langsdorfii)*. It differs from early blue violet by having larger flowers with shorter, thicker spurs. It has leafy stems, grows in wet soils, and is distributed from southeastern Alaska through Washington.

Early Blue Violet *Viola adunca*

Marsh Violet *Viola palustris*

Aleutian Violet *Viola langsdorfii*

Stream Violet *Viola glabella*

The most common violet in the region, this species grows from thick, spreading rhizomes. The stems, 6 to 12 inches (15 to 30 cm) long, are leafless on the lower half.

Leaves: Basal and alternate, or sometimes with opposite upper leaves; blades broadly heart shaped, toothed, and sharp pointed, 1 to 3 inches (3 to 7 cm) long and wide; petioles much longer, especially those of basal leaves.

Flowers: Bright yellow and showy, about ½ inch (10 to 15 mm) long; the lower one or three petals have purplish lines at the base.

Fruits: Many-seeded, explosive capsules.

Ecology: Inhabits roadside ditches, streams, and seepage areas in moist to wet subalpine meadows; distributed throughout the region.

A species that resembles the stream violet in general form and size is **Canada violet** *(V. canadensis),* but it has white flowers. It grows mainly in moist forests, often along streams, in Washington and southern British Columbia.

Round-Leaved Violet *Viola orbiculata*

Although this is a common plant, it usually goes unnoticed because it blooms early and the flowers are less showy than those of most violets. A perennial herb that spreads by rhizomes, its flowering stems are only 1 to 2 inches (2.5 to 5 cm) high.

Leaves: Mainly basal, sometimes persistent through the winter; blades round to broadly heart shaped, coarsely toothed, about 1½ inches (2 to 4 cm) long and as wide; more or less rounded at the tip; the petioles equal or exceed the blades in length.

Flowers: Pale yellow, ¼ to ½ inch (5 to 12 mm) long; the lower one or three petals have purplish lines.

Fruits: Many-seeded, explosive capsules.

Ecology: A plant of moist woods and subalpine meadows, distributed throughout the mountains of Washington and southern British Columbia.

A similar, less common but more conspicuous species is **evergreen violet** *(V. sempervirens).* It has shiny, thick, pointed, evergreen leaves marked by tiny purplish spots. Spreading by runners as well as rhizomes, it often forms dense mats and grows on the western slopes of the Cascades and Olympics in Washington and the Coast Mountains in southern British Columbia.

Stream Violet *Viola glabella*

Round-Leaved Violet *Viola orbiculata*

Evergreen Violet *Viola sempervirens*

Waterleaf Family

Hydrophyllaceae

A diverse family of annual and perennial herbs, the flowers are generally showy, radially symmetrical, and have parts in fives—five petals, fused to form a "cup" with five lobes; five sepals; and five stamens, derived from the petal cup and extended well beyond the petal lobes. In many species, the flowers are clustered into a dense spike or head that resembles a bottlebrush, with projected stamens for bristles. The leaves are usually pinnately compound, and the plants are usually hairy.

Ballhead Waterleaf

Hydrophyllum capitatum

One of the early flowering wildflowers, this is a low, succulent perennial herb with spreading stems and leaves and fleshy, fibrous roots.

Leaves: Alternate on short stems, pinnately compound with seven to eleven lobed leaflets; the petiole is grooved, allowing water to flow to the stem and downward to the fibrous roots, thus the name *waterleaf.*

Flowers: Blue to lavender-purple, showy, clustered into a ball-shaped head; stamens extended well beyond the petals.

Fruits: Capsules with one to three seeds.

Ecology: Ranges upward from the sagebrush steppe on the eastern side of the Cascades in Washington and adjacent British Columbia to the alpine zone; grows in moist thickets and open woods, on rocky slopes, and in other sites wet from melting snow.

Fendler's Waterleaf

Hydrophyllum fendleri

This is a hairy, succulent herb with thick rhizomes. The stems are solitary, weak, and stand up to 2 feet (0.6 m) high.

Leaves: Alternate, pinnately compound, with seven to thirteen irregularly toothed and sharp-pointed leaflets.

Flowers: White to pale lavender, about ⅜ inch (8 to 10 mm) long, several in loose clusters at the stem tips; sepals densely hairy.

Fruits: Capsules with one to three seeds.

Ecology: A common species in subalpine meadows throughout Washington and southern British Columbia.

At low to mid-elevations in moist woods on Vancouver Island and along the western slopes the Washington Olympics and Cascades, Fendler's waterleaf is replaced by **Pacific waterleaf** *(H. tenuipes),* a species with fewer leaflets and smaller, greenish white flowers.

Ballhead Waterleaf *Hydrophyllum capitatum*

Fendler's Waterleaf *Hydrophyllum fendleri*

Pacific Waterleaf *Hydrophyllum tenuipes*

Silky Phacelia

Phacelia sericea

A lovely plant, this perennial herb grows from a stout, branched root crown and has a cushionlike growth form. The stems are clustered and 4 to 12 inches (10 to 30 cm) high. Silky hairs clothe the entire plant (*sericea* means "silky").

Leaves: Basal and alternate, pinnately compound, the leaflets divided and toothed, giving the leaves the look of a fern.

Flowers: Bluish purple, densely clustered in an attractive terminal spike or head resembling a bottlebrush.

Fruits: Few-seeded capsules.

Ecology: Inhabits rocky, exposed areas, especially in the alpine zone; grows throughout the mountains of Washington and southern British Columbia.

White-Leaf Phacelia

Phacelia hastata

This is a hairy, perennial herb, usually with several stems derived from a thick, branched root crown. The species is highly variable–stems may be low and cushionlike or up to 2 feet (0.6 m) high.

Leaves: Basal and alternate, narrowly elliptical and undivided or with two basal, earlike lobes; blades up to 5 inches (12 cm) long, silvery hairy.

Flowers: White to pale blue, funnel shaped, borne in several dense, coiled clusters near the stem tips; stamens conspicuously extended beyond the petal tube.

Fruits: Few-seeded capsules.

Ecology: A common plant on the eastern slopes of the Cascades in Washington and adjacent British Columbia, where it ranges from the sagebrush steppe upward into the alpine zone; an occasional plant on rocky subalpine and alpine ridges west of the Cascade crest.

White-leaf phacelia is extremely variable, and several varieties have been recognized. The plants vary in size, growth form, extent of leaf lobing, and habitat. The plants are in the varieties ***leucophylla*** and ***leptosepala.*** Var. *leucophylla* is primarily a foothills species on the eastern side of the Cascades. Var. *leptosepala* is more cushionlike and grows in the high mountains. It is sometimes split off into a separate species, *P. leptosepala.*

Silky Phacelia
Phacelia sericea

White-Leaf Phacelia
Phacelia hastata var. *leptosepala*

Mistmaiden

Romanzoffia sitchensis

Mistmaiden is a delicate, perennial herb, often cushionlike. The stems are weak and typically arch downward.

Leaves: Mainly basal, with long, lax petioles and a roundish but coarsely toothed blade, ⅜ to 1½ inches (1 to 4 cm) wide.

Flowers: White and bell shaped, with a yellow center, about ¼ to ⅜ inch (6 to 10 mm) long and as wide, few to several borne on rather long, lax stalks along the upper half of the stems.

Fruits: Many-seeded capsules.

Ecology: An occasional plant on wet, gravelly slopes and ledges in the subalpine and alpine zones; distributed throughout the region.

This plant is often mistaken for one of the saxifrages, which have similarly lobed leaves and small, white flowers. It differs, however, in having fused petals with a yellow "eye."

Water-Lily Family

Nymphaeaceae

This is a family of aquatic plants, as the common name suggests, with large, attractive flowers and floating leaves. Only one species grows in the mountains of the Pacific Northwest.

Pond Lily

Nuphar polysepalum

A succulent, hairless, perennial herb, pond lily leaves and flowering stalks arise from thick, fleshy rhizomes embedded in mud at the bottom of shallow lakes and ponds.

Leaves: Large and heart shaped, 4 to 15 inches (10 to 38 cm) long and nearly as wide, floating on the water when fully developed and borne on ropelike petioles as long as the water is deep.

Flowers: Yellow to reddish tinged, waxy, large, and showy, about 3 inches (8 cm) across, solitary on long, ropelike stalks; sepals and petals several, merging in color and size; stamens numerous, reddish; pistil scalloped on top.

Fruits: Large many-seeded capsules.

Ecology: A common inhabitant of shallow lakes, ponds, and sluggish streams from lowlands to mid-elevations in the mountains; distributed throughout the region.

Mistmaiden *Romanzoffia sitchensis*

Pond Lily close-up *Nuphar polysepalum*

Pond Lily habitat *Nuphar polysepalum*

WILLOW FAMILY

Salicaceae

The willow family is made up of trees and shrubs with alternate, simple leaves. The plants are unisexual, bearing either male or female flowers in catkins. Each flower is associated with a scale or bract, which is egg shaped and smooth in willows and frilly edged in cottonwoods and aspens. The individual flowers are small, highly reduced, and inconspicuous. In addition to lacking one or the other of the reproductive organs (stamens or pistil[s]), the flowers have no sepals or petals. This condition of reduced, densely congested, unisexual flowers is an adaptation to wind pollination. Willow catkins are visited by bees, however, which collect both pollen and nectar and undoubtedly are important pollinators.

Trees in the willow family are treated in the Trees section (pages 92 to 95); descriptions of other plants in the willow family (*Salix* ssp.) are found here.

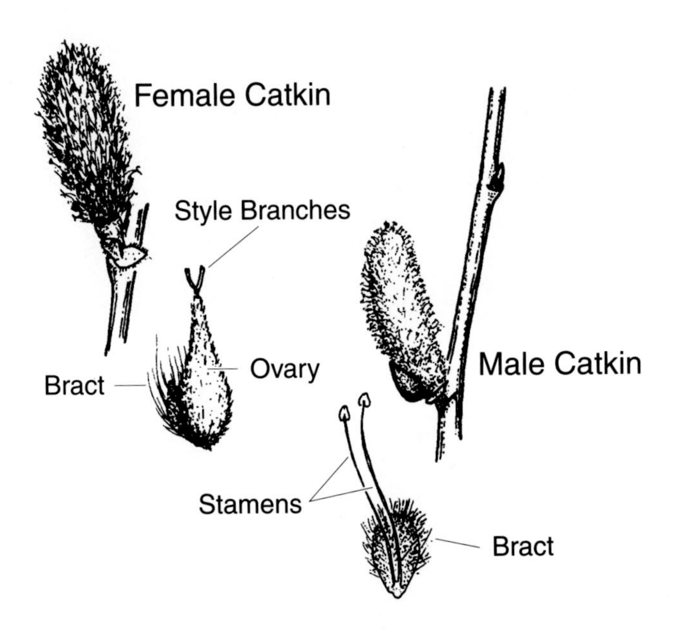

Cascade Willow in autumn *Salix cascadensis*

Arctic Willow

Salix arctica

A dwarf shrub, arctic willow is usually prostrate or trailing but sometimes (at lower elevations) up to 20 inches (50 cm) tall. The branches are brown and stout; the twigs vary from sparsely hairy to hairless.

Leaves: Broadly elliptical to nearly round, sparsely hairy on the undersurface, often with a tuft of hair at the tip; apex rounded to pointed; leaf margins smooth, nontoothed.

Flowers: Catkins 1 to 2 inches (2 to 5 cm) long, borne on prominent, leafy branchlets; pistils hairy; styles about 0.6 to 2.2 mm long, reddish, drying to purplish; floral bracts brown to blackish and densely hairy.

Fruits: Capsules, covered with spreading hairs.

Ecology: A common species in Alaska and British Columbia, less so in Washington; prefers moist sites in the subalpine and alpine zones, where it is often a community dominant.

Arctic willow could be mistaken for **stoloniferous willow** *(S. stolonifera)* in the northern part of the range (north of 55 degrees). Unlike arctic willow, stoloniferous willow has capsules that are hairless or have only a few hairs at the tip.

Cascade Willow

Salix cascadensis

This dwarf, mat-forming shrub is never more than a few inches high. The branches are mainly gray and hairless.

Leaves: Narrowly elliptical, with a pointed tip, ⅜ to ¾ inch (1 to 2 cm) long, hairy (at least when young); margins smooth, nontoothed.

Flowers: Catkins up to 1 inch (2.5 cm) long, borne on leafy branchlets; pistils hairy; styles 1 to 2 mm long; floral bracts brown to blackish and densely hairy.

Fruits: Capsules, covered with spreading hair.

Ecology: Common willow in the upper subalpine and alpine zones of the Washington Cascades, growing in dry meadows and along ridges; ranges northward into southern British Columbia and southward into Oregon.

Cascade willow resembles arctic willow but has smaller, narrower leaves and smaller catkins. The two species reputably hybridize where their ranges overlap.

Arctic Willow (male) *Salix arctica*

Arctic Willow (female) *Salix arctica*

Cascade Willow *Salix cascadensis*

Variable Willow *Salix commutata*

This willow is a medium-sized shrub, up to 10 feet (3 m) tall. The branches are dark brown, densely covered with woolly hairs when young but usually becoming hairless with age.

Leaves: Broadly egg shaped, 1½ to 3 inches (4 to 8 cm) long, pointed at the tip, hairy on both sides; margins smooth to finely toothed.

Flowers: Catkins, sometimes more than 2 inches (2 to 6 cm) long, borne on leafy shoots, appearing with the leaves or after they have developed; pistils hairy; styles about ¼ inch (0.5 to 1.2 mm) long; floral bracts brown.

Fruits: Glabrous (hairless) capsules on a stalk covered with silky hairs.

Ecology: Moist to wet streamsides, lakeshores, open forests, gravelly river terraces, and glacial moraines in the montane and subalpine zones; a common and locally dominant species throughout the Pacific Northwest.

Gray-Leaved Willow *Salix glauca*

Gray-leaved willow, another medium-sized shrub, grows to 10 feet (3 m) tall, with reddish brown to grayish branches. The young twigs are densely to sparsely hairy, becoming less so with age.

Leaves: Elliptical to egg shaped, 1 to 2 inches (2.5 to 5 cm) long, hairy on both sides and whitish on the underside; leaf margins smooth and nontoothed.

Flowers: Catkins up to 2 inches (5 cm) long, borne on short, leafy shoots; pistils hairy; styles tiny (0.5 to 0.8 mm); floral bracts greenish, drying brown.

Fruits: Sparsely hairy capsules, stalks covered with silky hairs.

Ecology: Moist to wet streamsides and lakeshores, moist to dry forests, and avalanche tracts in the montane to lower alpine zones; a common, often dominant species throughout the northern half of the region.

Gray-leaved willow is a variable species, often difficult to distinguish from other medium-sized willows such as *S. commutata,* described above, and **Barclay's willow** *(S. barclayi).* The separation of the three species is based on a combination of characteristics to include nonglaucous (whitish) leaves in *S. commutata* and nonhairy (glabrous) pistils in *S. barclayi.* The three species grow in similar habitats and may hybridize where their ranges overlap.

Variable Willow *Salix commutata*

Gray-Leaved Willow *Salix glauca*

Barclay's Willow *Salix barclayi*

Net-Veined Willow

Salix reticulata ssp. *reticulata*

A dwarf, usually prostrate shrub, this willow is no more than 4 inches (1 to 10 cm) high, often rooting along the stems and forming small mats. The branches are green to greenish brown and hairless.

Leaves: Broadly elliptical or egg shaped to round, up to 2 inches (0.5 to 5 cm) long, dark green and leathery, conspicuously net veined; apex rounded; margins nontoothed or sometimes with minute teeth, rolled under.

Flowers: Catkins about 1 to 2 inches (2 to 6 cm) long, with an equally long stalk; pistils hairless or more commonly sparsely to densely silky; styles tiny (0.2 to 1.0 mm); floral bracts greenish, reddish, or brown, hairy.

Fruits: Capsules, red, hairless or with silky hairs.

Ecology: A conspicuous and common willow in the subalpine and alpine zones of Alaska and northern British Columbia (north of 54 degrees).

In the southern part of the region, particularly in Washington, net-veined willow is replaced by **dwarf snow willow** (*S. reticulata* ssp. *nivalis*). This attractive dwarf subspecies is easily distinguished from the more northern subspecies by a lack of prominent veins on the upper surface of the leaves. The leaves are also smaller. **Glabrous dwarf willow,** another subspecies *(glabellicarpa),* grows in the region too but is restricted to the Queen Charlotte Islands and extreme southeastern Alaska. This subspecies is distinctive in having glabrous (nonhairy) pistils and capsules.

Net-Veined Willow *Salix reticulata* ssp. *reticulata*

Dwarf Snow Willow *Salix reticulata* ssp. *nivalis*

Dwarf Snow Willow autumn colors *Salix reticulata* ssp. *nivalis*

Sitka Willow

Salix sitchensis

A tall shrub or small tree, Sitka willow grows to 30 feet (8 m) tall. The branches are dark brown to gray and sparsely hairy. Velvety hair covers the young twigs, which are notably brittle.

Leaves: Broadly lance shaped to elliptical, about 1 to 3 inches (2 to 9 cm) long, margins smooth or dotted with glands, upper surface dark green and sparsely silky, undersurface densely covered with silky gray hair, contrasting in color with the upper side.

Flowers: Catkins 1 to 4 inches (2.5 to 10 cm) long, borne on short branchlets; pistils covered with silky hairs; styles tiny (0.4 to 0.8 mm); floral bracts brown and densely hairy.

Fruits: Capsules with silky hairs.

Ecology: Moist to wet streamsides, lakeshores, forest margins, and avalanche tracts in the montane zone; a common species throughout the region.

When it is flowering, Sitka willow is fairly distinct from other *Salix* species. It is unusual in having a single stamen per male flower, and the female catkins are long, narrow, and typically arched. Specimens lacking catkins are sometimes confused with **Scouler's willow** *(S. scouleriana).* The major distinguishing vegetative characteristic is the silky hair on the undersurface of Sitka willow leaves, which contrasts sharply with the upper leaf surface. Also, Scouler's willow is often much larger and more treelike, with less-brittle twigs.

Scouler's Willow

Salix scouleriana

A common Pacific Northwest species, Scouler's willow is a tall shrub or medium-sized tree, up to 50 feet (15 m) tall. The branches are reddish brown and glossy, and the young twigs are densely covered with velvety hairs.

Leaves: Elliptical to egg shaped, 2 to 5 inches (3 to 12 cm) long, dark green above, sparsely to densely hairy and whitish on the undersurface; margins nontoothed to coarsely toothed.

Flowers: Catkins 1 to 3 inches (2.5 to 7.5 cm) long, borne on the previous year's branches, usually before new leaves appear; pistils densely covered with silky hair; styles tiny (0.2 to 0.8 mm); floral bracts dark brown to black.

Fruits: Capsules with silky hairs.

Ecology: A frequent species throughout most of the Pacific Northwest. Distributed in open forests and along moist riversides and lakeshores, ranging from lowlands up to the subalpine zone.

Sitka Willow (female) ***Salix sitchensis***

Sitka Willow (male) ***Salix sitchensis***

Scouler's Willow (female) ***Salix scouleriana***

GRAMINOIDS

Graminoids include the grass, sedge, and rush families, which share grass-like characteristics, including long, narrow, swordlike leaves with parallel veins and thin, tough stems that bear clusters of inconspicuous flowers. The reduced flowers typify the wind pollination strategy so highly developed by the graminoids. Although members of these families superficially resemble each other and are frequently regarded collectively as "grasses," they are only distantly related. The differences and similarities in the three families are presented in the accompanying table. Some of the plant parts with technical names are shown on the grass, sedge, and rush illustrations.

	GRASSES	SEDGES	RUSHES
LEAVES	(Poaceae) Mainly flat and swordlike; blades extend from sheaths that wrap around the stem but usually remain open, at least at the top (sheath edges unsealed).	(Cyperaceae) Mainly flat and swordlike; blades extend from closed sheaths that wrap around the stem (sheath edges fused).	(Juncaceae) Flat and swordlike or rounded, looking like a continuation of the stem; sheaths not obvious.
STEMS	Round or somewhat flattened, hollow except at the conspicuous, swollen nodes.	Mainly triangular, with three rows of leaves, becoming hollow only with age; nodes not swollen or conspicuous.	Round and solid, not hollow; nodes not swollen or conspicuous.
INFLORESCENCE	A panicle or spike; flowers compressed into spikelets with glumes, lemmas, and paleae; flowers usually bisexual and lacking petals and sepals.	One or more densely compressed spikes; flowers unisexual, lacking sepals and petals, and each is associated with a single bract; male flowers reduced to three stamens; female flowers reduced to an ovary enclosed in a saclike perigynium.	A panicle or spike; flowers bisexual, with sepals and petals, although they are reduced, bractlike, and inconspicuous.
FRUITS	A grain, with the ovary wall fused to and seemingly a part of the single seed.	An achene, the ovary wall not fused to the single seed.	A capsule, with three *(Luzula)* or several *(Juncus)* seeds.

Grass Family

Poaceae (Gramineae)

The grass family is one of the largest and most successful plant families and is well represented in the mountains of the Northwest, particularly in the subalpine and alpine zones. In general, grasses tolerate considerable drought and are typically among the dominant plants on dry ridges and open south-facing slopes, notably on the eastern side of the mountain ranges and in the sagebrush steppe below. The ecological importance of the family is magnified by its valuable role as a food crop for grazing animals and seed "predators," including the large population of humans that depends on cereal crops for sustenance.

Grasses are attractive, but they lack the colorful flowers that characterize plants pollinated by insects and other animals. Their flowers are individually very small and highly reduced–a specialization for wind pollination. They lack sepals and petals; the anthers are large and produce massive amounts of pollen; and the style branches are long and feathery, suited to combing the air for pollen. Both the anthers and style branches extend well beyond the associated bracts (glumes, lemmas, and paleae) when the flowers are mature. Typically, the pollen is released before or after the styles are receptive, thus preventing self-pollination of the bisexual flowers.

Because they do not have showy flowers, grasses are distinguished by a combination of spikelet and vegetative characteristics.

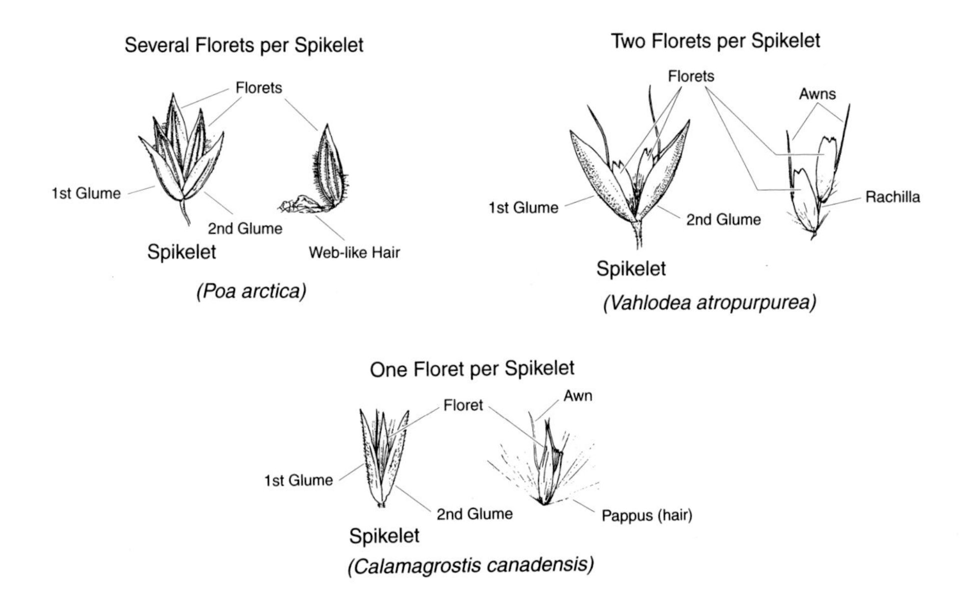

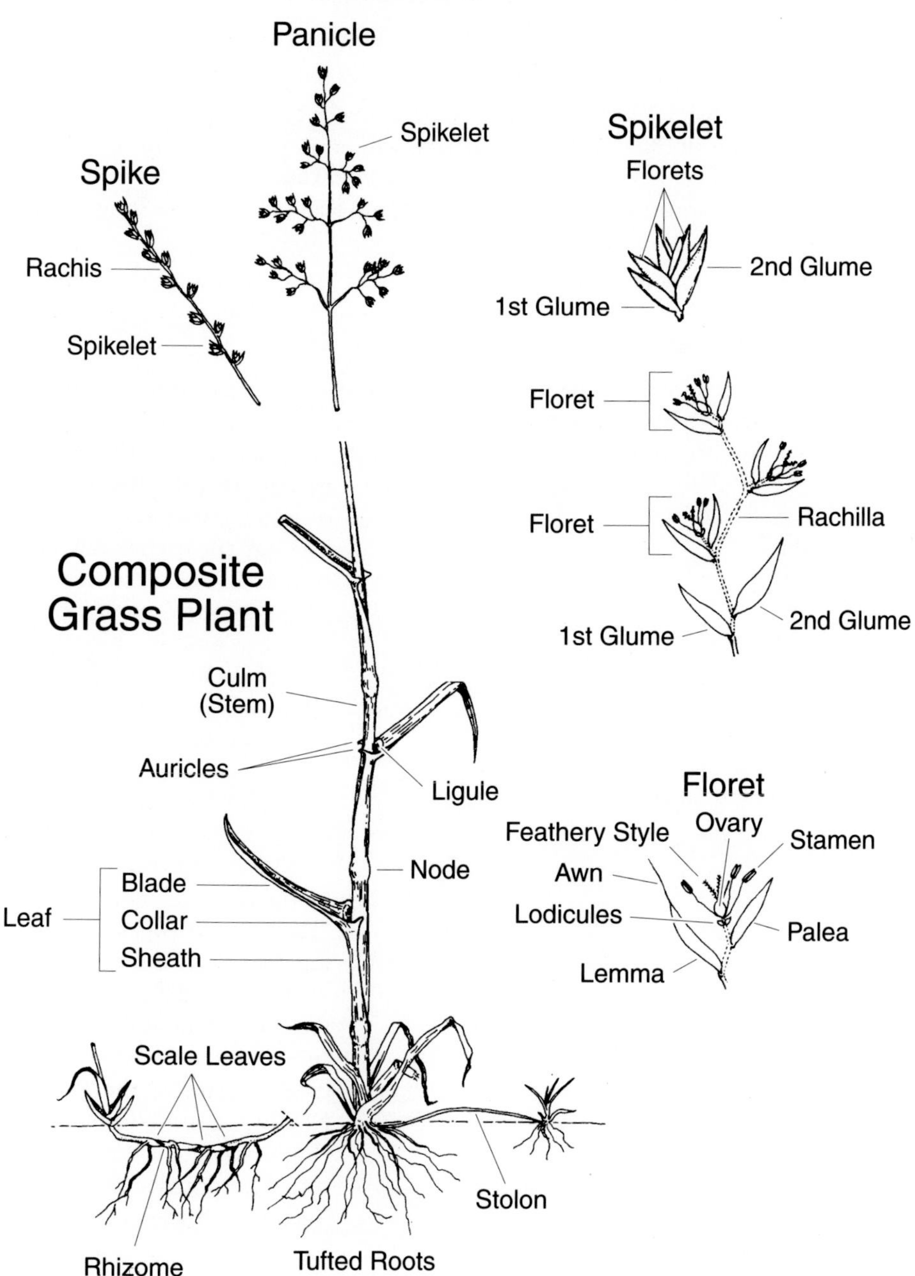
Poaceae
(Grasses)
Panicle
Spikelet
Spike
Rachis
Spikelet
Spikelet
Florets
2nd Glume
1st Glume
Floret
Floret
Rachilla
2nd Glume
1st Glume
Composite
Grass Plant
Culm
(Stem)
Auricles
Ligule
Node
Leaf
Blade
Collar
Sheath
Floret
Feathery Style
Ovary
Stamen
Awn
Lodicules
Palea
Lemma
Scale Leaves
Stolon
Rhizome
Tufted Roots

Bluejoint Reedgrass *Calamagrostis canadensis*

This coarse, tufted perennial may be as much as 3 feet (1 m) tall. The plants are hairless or nearly so and spread by rhizomes.

Leaves: Long, lax, and flat, about ¼ inch (5 to 10 mm) wide; the blade lacks auricles, and the ligules are up to ⅜ inch (1 cm) long.

Inflorescence: A large but narrow, nodding or arching purplish panicle, up to 1 foot (30 cm) long; spikelets, with a single floret; glumes about ¼ inch (4 to 6 mm) long; lemma bears a delicate awn arising from near the base and surrounded by bristly hairs that are about as long as the lemma.

Ecology: Prefers moist to wet meadows and forest openings; distributed throughout the Pacific Northwest, ranging from low elevations to the subalpine zone.

Bluejoint reedgrass is extremely variable, and several taxonomic varieties have been recognized. Other species of *Calamagrostis* grow in the region, too, but none are as robust as bluejoint reedgrass. One species that often grows as an understory dominant in montane forests on the drier, eastern slopes of the mountains is **pinegrass** *(C. rubescens),* which resembles bluejoint reedgrass but is smaller in all respects and has hairs much shorter at the base of the lemma. The two species are found in distinctly different habitats.

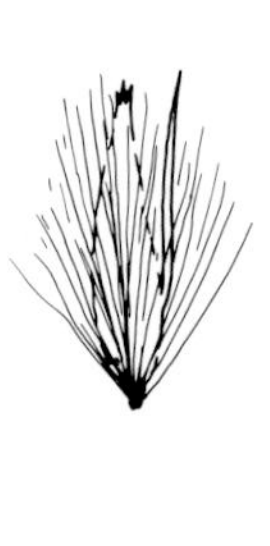

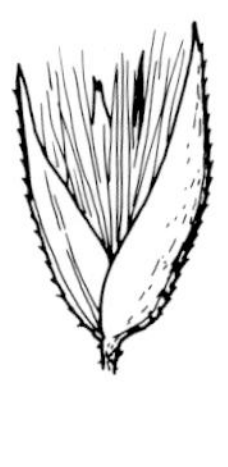

The bentgrasses (genus *Agrostis*) are similar to *Calamagrostis* species in having a single floret per spikelet, but they differ in lacking the tuft of hair at the base of the lemma. Several bentgrasses grow in the Pacific Northwest, usually on alpine ridges. Most of them are low, tufted perennials with open, purplish panicles.

Bluejoint Reedgrass *Calamagrostis canadensis*

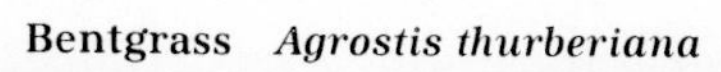

Bentgrass *Agrostis thurberiana*

Timber Oatgrass

Danthonia intermedia

A densely tufted perennial grass, this species varies in height from 3 inches to 2 feet (7 to 60 cm).

Leaves: Mainly basal, usually folded or rolled but sometimes flat, about ⅛ inch (3 mm) wide; usually hairy on the undersurface, particularly at the base of the blade.

Inflorescence: A densely compacted panicle, 1 to 2 inches (2 to 5 cm) long; spikelets large, with three or more florets; glumes about ⅝ inch (15 mm) long; lemmas hairy at the base and along the margins, about ⅜ inch (7 to 10 mm) long and deeply toothed at the tip, with a twisted awn borne between the two teeth.

Ecology: Found on rocky ridges and in rather dry meadows, from lowlands into the alpine zone; distributed throughout the region, but most common in Washington and southern British Columbia.

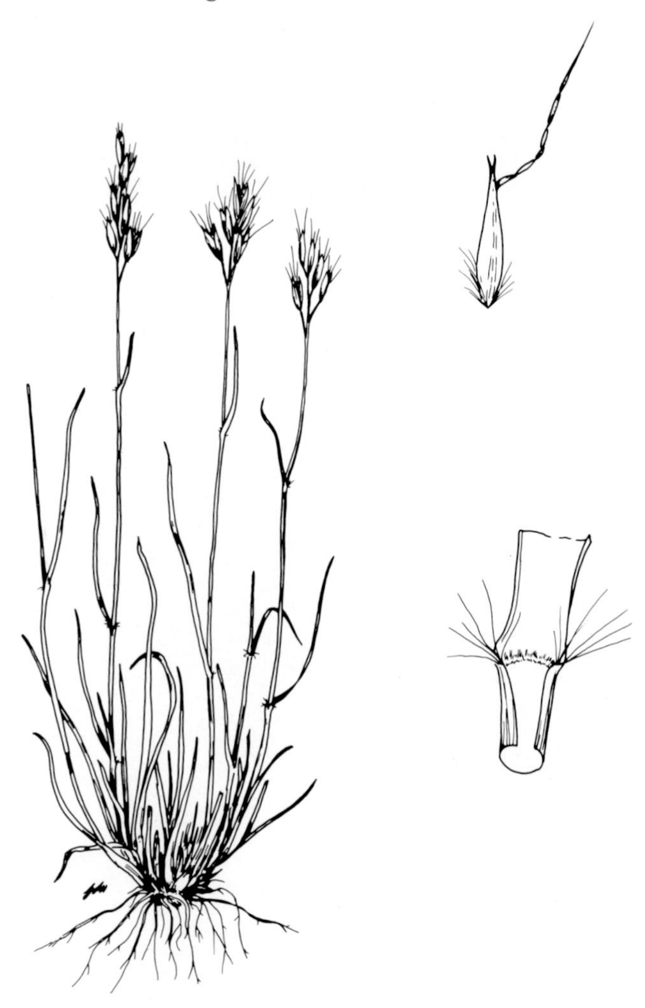

Timber Oatgrass *Danthonia intermedia*

Blue Wildrye

Elymus glaucus

Blue wildrye is a tall, tufted perennial grass with fibrous roots. The stems are often more than 3 feet (1 m) tall.

Leaves: Usually flat and lax, ⅛ to ⅝ inch (3 to 15 mm) wide, sometimes with rough hairs; auricles at the base of the blade well developed; ligules tiny (0.5 to 1.5 mm).

Inflorescence: A dense spike up to 6 inches (15 cm) long; spikelets paired, two per node, with three to five florets; glumes narrowly lance shaped, about ⅝ inch (15 mm) long; lemmas somewhat shorter than the glumes and usually awned from the tip.

Ecology: A fairly common grass in open forests and dry meadows, usually in rocky soils, ranging from the southern panhandle of Alaska southward throughout the region.

Blue Wildrye *Elymus glaucus*

Idaho Fescue

Festuca idahoensis

A species with several close look-alikes, this is a densely tufted perennial with fibrous roots. The stems are up to 3 feet (0.9 m) tall, much longer than the numerous leaves. The entire plant is bluish.

Leaves: Mainly basal and densely clustered; blades very narrow, folded or rolled, usually less than 6 inches (15 cm) long.

Inflorescence: A narrow, erect panicle up to 7 inches (18 cm) long; spikelets about ½ inch (12 mm) long, with five to seven florets; lemmas longer than the glumes and awned from the tip.

Ecology: Often the major dominant species in lowland forest understories on the eastern side of the mountain ranges; less common in dry meadows and on rocky montane ridges toward the west; distributed from southern British Columbia through Washington.

Toward the north, Idaho fescue is replaced by **Rocky Mountain fescue** *(F. saximontana),* a similar grass with smaller spikelets and a narrower, more dense panicle. It is distributed throughout the Pacific Northwest in dry forest openings and on rocky slopes up to the alpine zone. In moist, subalpine and alpine meadows, especially in the Washington Cascades, **green fescue** *(F. viridula)* is often a dominant species. It is more robust than Idaho fescue and, as the common and Latin names suggest, is much greener. On dry, rocky alpine ridges, **alpine fescue** *(F. brachyphylla)* is a common grass, ranging from Alaska through Washington. This fescue is a dense cushion plant, usually not more than about 6 inches (15 cm) high, and appears to be a miniature form of Rocky Mountain fescue; the two taxa have previously been combined as varieties in the taxon *F. ovina.*

Idaho Fescue *Festuca idahoensis*

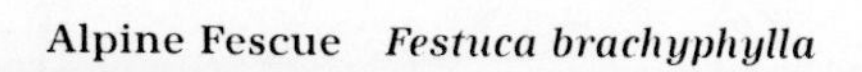

Alpine Fescue *Festuca brachyphylla*

Alpine Timothy

Phleum alpinum

This grass is a tufted perennial with fibrous roots. The stems are 5 to 24 inches (12 to 65 cm) tall and sometimes root at the lower nodes.

Leaves: Flat, ⅛ to ⅜ inch (2 to 9 mm) wide, rough on the margins; ligules about ⅛ inch (1 to 4 mm) long.

Inflorescence: A very dense, spikelike, cylindrical panicle, ⅜ to 2 inches (1 to 5 cm) long and ⅜ inch (1 cm) wide; spikelets bear a single floret; glumes fringed with hair on the keel and usually hairy on the sides, abruptly narrowing to a stout ⅛-inch (3 mm) awn; lemmas with minute hairs.

Ecology: A common montane to alpine grass found in moist forest openings, along stream banks, and in meadows throughout the region.

Timothy *(P. pratense),* an introduced species, may grow in the same montane zone, usually along roads and trails, and near pastures where commercial hay has provided a seed source. It differs from alpine timothy in having a much longer panicle and taller stems.

Alpine Timothy ***Phleum alpinum***

Alpine Bluegrass *Poa alpina*

A tufted plant with fibrous roots, its stems are 2 to 24 inches (5 to 60 cm) tall.

Leaves: Flat and relatively broad, about ⅛ inch (2 to 4 mm) long, mainly basal; ligules about ⅛ inch (1 to 4 mm) long.

Inflorescence: A compact, pyramidal panicle, ¾ to 3 inches (2 to 8 cm) long; spikelets relatively large, with three to six florets; glumes broad, with short hairs above; lemmas coarsely hairy on the keel and margins, unawned.

Ecology: Common throughout the Pacific Northwest in meadows and on rocky slopes in the subalpine and alpine zones.

Alpine bluegrass, with distinctive broad basal leaves, is one of the few easily identified members of this difficult genus. Nevertheless, **Cusick's bluegrass** (*P. cusickii* ssp. *epilis*) is often mistaken for alpine bluegrass. It grows in similar habitats in the southern part of the region and can be distinguished by its narrower, often rolled leaves and more compact, egg-shaped panicle.

Arctic bluegrass *(P. arctica)* also resembles alpine bluegrass but differs in having spreading rhizomes and cobwebby hairs at the base of the lemmas. It is common in southeastern Alaska and British Columbia, except along the outer Coast Mountains. It grows in meadows, on moist rocky slopes, and along streams in the subalpine and alpine zones.

Alpine Bluegrass *Poa alpina*

Cusick's Bluegrass *Poa cusickii* ssp. *epilis*

Spike Trisetum

Trisetum spicatum

A densely tufted perennial grass, with stems 3 to 20 inches (10 to 50 cm) tall.

Leaves: Folded to flat, about ⅛ inch (2 to 5 mm) wide, covered with fine hairs.

Inflorescence: A dense, spikelike, cylindrical panicle, 2 to 6 inches (5 to 15 cm) long, often turning purple with maturity; spikelets bear two florets; glumes nearly equal in size; lemmas with a long, bent awn from the middle of the back, making the inflorescence look bristly.

Ecology: A common montane to alpine grass found on dry, rocky slopes and rock outcrops throughout the region.

Spike Trisetum *Trisetum spicatum*

Mountain Hairgrass *Vahlodea (Deschampsia) atropurpurea*

This species is a loosely to densely tufted perennial with rather weak, spreading stems 8 to 30 inches (20 to 75 cm) long. The species name, *atropurpurea,* relates to the dark purplish color of the spikelets.

Leaves: Lax and flat, about ¼ inch (4 to 6 mm) wide; usually hairless; ligules about ⅛ inch (1.5 to 3.5 mm) long.

Inflorescence: A loose, nodding panicle; spikelets with two, rarely three, florets, conspicuously purple; glumes about ¼ inch (5 mm) long and nearly equal; lemmas shorter, surrounded by stiff hairs, and bearing a stout, twisted, bent awn from below the middle.

Ecology: A common grass on moist, subalpine slopes and along streams, particularly in areas of late snowmelt; distributed throughout the region.

Sedge Family Cyperaceae

The sedge family is large and composed of several genera. In some genera, sepals and petals are modified into capillary or cottony bristles. The largest genus, *Carex,* has sepals and petals replaced by a perigynium, a saclike structure that encloses the ovary. The flowers are bisexual in some genera and unisexual in others. With a few notable exceptions, most members of the family are aquatics, or at least prefer wet habitats.

In Pacific Northwest mountains, only two genera are well represented, except in some of the lowland lakes and ponds. These are the sedges *(Carex)* and the cotton-grasses *(Eriophorum).* Cotton-grasses are easily identified by their long, cottony bristles, which are modified sepals and/or petals. Except for the inflorescence, cotton-grasses resemble sedges.

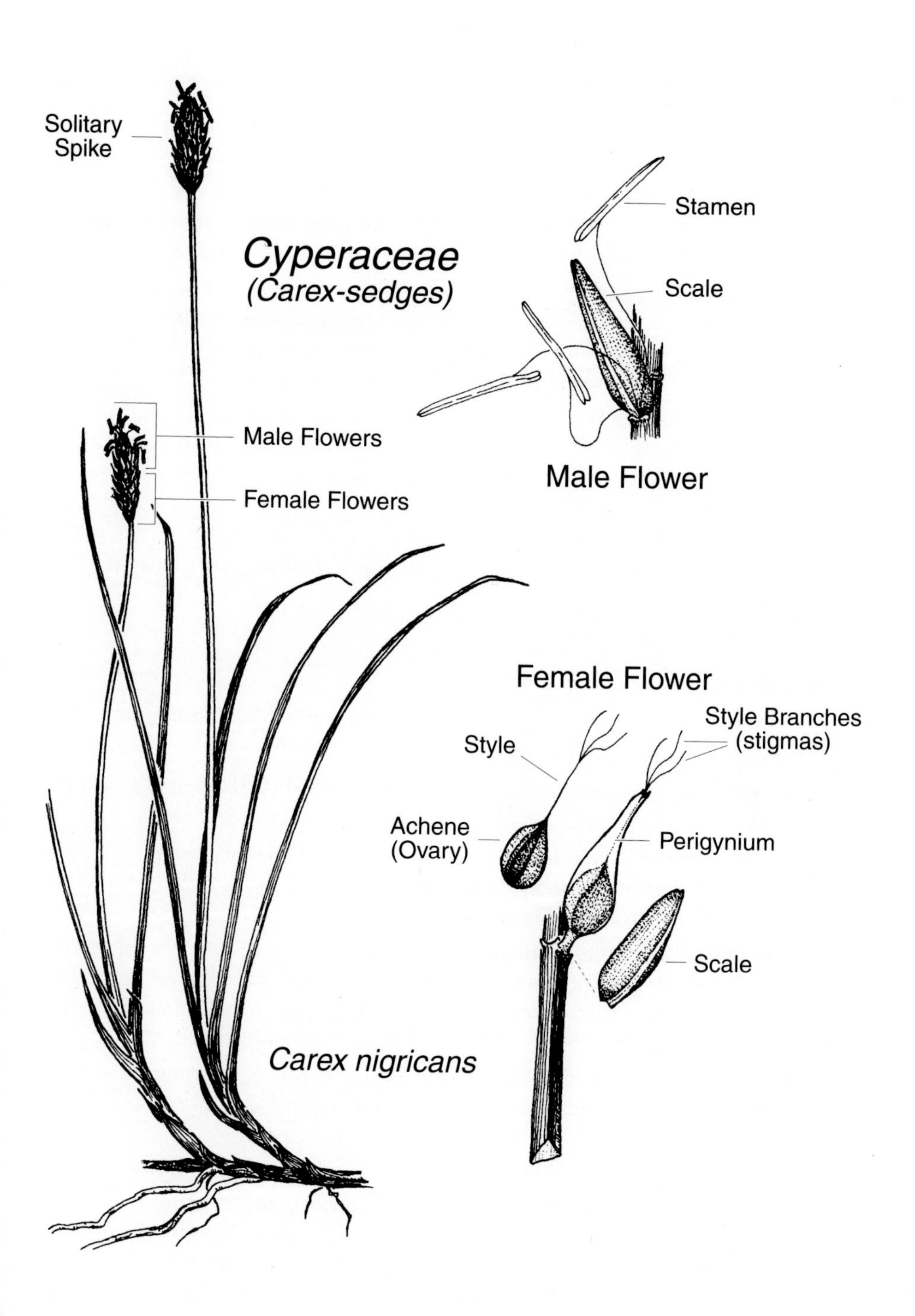
Solitary
Spike
Cyperaceae
(Carex-sedges)
Stamen
Scale
Male Flowers
Female Flowers
Male Flower
Female Flower
Style Branches
(stigmas)
Style
Achene
(Ovary)
Perigynium
Scale
Carex nigricans

Two-Toned Sedge *Carex albonigra*

This loosely tufted sedge has short rhizomes and stiff stems 4 to 12 inches (10 to 30 cm) high. Leaves from previous seasons are more or less persistent at the base of the stem.

Leaves: Numerous, flat, usually less than 4 inches (10 cm) long, and about ¼ inch (2 to 7 mm) wide, clustered at the base of the stem.

Inflorescence: Spikes, usually two to four, the terminal one about ⅜ inch (1 to 2 cm) long, with female flowers above the male flowers; the other spikes are smaller and have female flowers only; the bract below the lowermost spike is as long as the inflorescence.

Perigynia: Dark reddish brown or blackish purple, about ⅛ inch (2.7 to 3.6 mm) long, with a short beak; scales dark like the perigynia but with pale margins (two-toned), elliptical to round, and about as long as the perigynia; stigmas three.

Ecology: A sedge of dry meadows, ridges, and rocky slopes in the subalpine and alpine zones; common in the southern part of the region, becoming less frequent northward.

Showy Sedge *Carex spectabilis*

This sedge is more or less tufted, with short, stout, scaly rhizomes. The stems are slender and 6 to 36 inches (15 to 90 cm) tall.

Leaves: Flat, 2 to 6 inches (5 to 20 cm) long, and about ¼ inch (2 to 7 mm) wide.

Inflorescence: Three to six spikes, usually ⅜ to 1¼ inches (1 to 3 cm) long, the uppermost with only male flowers; the leaflike bract subtending the lowest spike is about as long as the inflorescence.

Perigynia: Pale green, purplish or brown, elliptical, not ribbed, about ⅛ to ¼ inch (3 to 5 mm) long, with a slender, toothed beak at the tip; female scales with a conspicuous midvein, usually longer then the perigynia, dark reddish brown to blackish; stigmas three.

Ecology: Occupies moist sites from montane forest openings to subalpine and alpine meadows; common throughout the region, especially toward the south.

Short-stalked sedge *(C. podocarpa)* closely resembles showy sedge but has two-ribbed perigynia, and the female scales lack a prominent midvein. It is common in Alaska and northern British Columbia. **Fragile sedge** *(C. membranacea)* also resembles showy sedge but has stiffer stems with old, persistent leaves at their base, a somewhat inflated perigynia without prominent teeth at the tip, and inflorescence scales lacking a prominent midvein. This sedge is common in bogs, wet meadows, and lakeshores in northern British Columbia and Alaska.

Two-Toned Sedge *Carex albonigra*

Showy Sedge *Carex spectabilis*

Black Alpine Sedge *Carex nigricans*

The stems of this sedge are stiffly erect, 2 to 14 inches (5 to 35 cm) high, usually in small clusters, and borne on short rhizomes. Old leaves persist at the base of the stems.

Leaves: Numerous and stiff, flat or slightly channeled, usually shorter than the stems, 1½ to 5 inches (4 to 13 cm) long and ⅜ to 1¼ inches (1 to 3 cm) wide.

Inflorescence: Solitary spikes, male flowers above the female flowers, bractless below the spike.

Perigynia: Brownish, narrowly elliptical to lance shaped, about ⅛ to ¼ inch (3 to 4.5 mm) long with a basal stalk and a short beak at the tip, spreading to reflexed backward at maturity and soon deciduous; female scales equal to or shorter than the perigynia, dark brown or blackish; stigmas three.

Ecology: A subalpine and alpine sedge common throughout the region, particularly in areas with late snowmelt. Also distributed in wet meadows, along streams, and near ponds and lakes.

Black alpine sedge could be mistaken for **Pyrenean sedge** *(C. pyrenaica)*, which is another widespread, high-elevation sedge, but with no rhizomes and much narrower leaves–less than ⅛ inch (0.5 to 1.5 mm) wide.

Several-Flowered Sedge *Carex pluriflora*

A northern sedge, this species has solitary or loosely clustered stems growing from purplish black rhizomes. The stems are 6 to 24 inches (20 to 75 cm) high.

Leaves: Flat and narrow, about ⅛ inch (2 to 4 mm) broad, and about as long as the stems.

Inflorescence: Two or three spikes, the upper one erect and male, the lower one to two female and drooping on threadlike stalks; the bract below the lowest spike is bristlelike and much shorter than the inflorescence.

Perigynia: Blackish to dark brown, egg shaped, about ⅛ inch (3 to 4.5 mm) long, rather plump and lacking a beak; scales black and slightly longer than the perigynia; stigmas three.

Ecology: A common sedge in southeastern Alaska and the Coast Mountains of British Columbia, becoming rare in the North Cascades of Washington; found in bogs, wet meadows, and along lakeshores from lowlands to mid-elevations in the mountains.

Black Alpine Sedge *Carex nigricans*

Several-Flowered Sedge *Carex pluriflora*

Mertens' Sedge *Carex mertensii*

Mertens' sedge is a robust, rhizomatous sedge with solitary or loosely clustered stems 16 to 40 inches (40 to 120 cm) tall.

Leaves: Numerous, large, flat, ⅛ to ⅜ inch (3 to 10 mm) wide, located mainly on the stem.

Inflorescence: Five to ten spikes an inch or more (2 to 4 cm) long, nodding on slender stalks; on the terminal spike, female flowers are above the male flowers; for the most part, the lower spikes have female flowers only.

Perigynia: Pale, with numerous small, brown or reddish brown dots, about ¼ inch (4 to 5.4 mm) long, broadly elliptical to round, with a very short beak; female scales dark brown to purplish black, with a pale midrib and a short, spiny tip; all but the lower scales are much shorter than the perigynia; stigmas three.

Ecology: Common in ditches, moist meadows, and along streams in the montane zone throughout the region.

Dunhead Sedge *Carex phaeocephala*

Dense tufts characterize this sedge, which has slender stems 2 to 16 inches (5 to 40 cm) high.

Leaves: Numerous, clustered at the base of the stems, stiff and flat but narrow, less than ⅛ inch (0.5 to 2.5 mm) wide.

Inflorescence: Three to seven densely clustered spikes in a stiff, straw-colored head; male flowers are below the female flowers in each spike.

Perigynia: Tan or brown, up to ¼ inch (6 mm) long, abruptly tapering into a short beak; scales about as long and wide as the perigynia; margins of the perigynia very thin (winged); stigmas two.

Ecology: A common subalpine-alpine species throughout the Pacific Northwest with the exception of the Queen Charlotte Islands; found on moderately dry to dry rocky slopes and meadows.

Several other species resemble dunhead sedge, but they usually have darker, not straw-colored, spikes. Also, the back side of the perigynia in dunhead sedge has several conspicuous veins. The most common look-alike is probably **Falkland Island sedge** (*C. macloviana,* to include *C. pachystachya* and *C. microptera).* It can be distinguished from dunhead sedge by its darker spike head, which is copper colored to dark brown. Also, it has scales smaller than the perigynia. A highly variable and taxonomically difficult species, Falkland Island sedge grows in a variety of habitats–moist meadows, forest openings, and dry grassy slopes, ranging from lowlands to the subalpine zone.

Mertens' Sedge *Carex mertensii*

Dunhead Sedge *Carex phaeocephala*

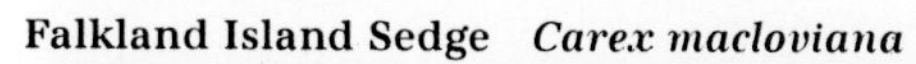

Falkland Island Sedge *Carex macloviana*

Scheuchzer's Cotton-Grass *Eriophorum scheuchzeri*

An attractive wetland plant, this species spreads by an extensive rhizome system and forms large clonal populations. The stems are slender and 4 to 16 inches (10 to 40 cm) tall.

Leaves: Few, channeled or triangular, very narrow (only about 1 mm wide), borne near the base of the stem and shorter than the stem.

Inflorescence: Solitary, terminal spike up to 1 inch (0.8 to 2 cm) long, upper flowers male, and the lower ones bisexual; each flower has numerous white, cottony bristles.

Fruits: Brown or blackish, about ⅝ inch (1.7 to 2.3 mm) long, broadly lance-shaped, triangular achenes, each surrounded by numerous silky white bristles.

Ecology: A plant of wet meadows and bogs, riverbanks, and lakeshores from the montane to alpine zones; common in the northern half of the region.

Other species of solitary-spiked cotton-grass grow in the same habitat and locality, but they either lack rhizomes *(E. brachyantherum)* or are more robust and have rust-colored bristles attached to the base of the achene *(E. chamissonis).*

Narrow-Leaved Cotton-Grass *Eriophorum angustifolium*

The plant stems in this species are roundish and solitary or a few together, arising from creeping rhizomes. The plants grow to about 3 feet (1 m) tall, and the stem base is clothed with brown leaf sheaths. (This species is often referred to as *E. polystachion.*)

Leaves: Lower leaves flat, about ¼ inch (4 to 6 mm) wide; upper leaves channeled, triangular, and narrower.

Inflorescence: Two to several drooping spikelets in a loose cluster; two or more leafy bracts mark the base of the inflorescence; flowers bisexual.

Fruits: Dark brown to black, egg-shaped, triangular achenes, each surrounded by numerous long silky white bristles.

Ecology: A common inhabitant of bogs and wet meadows throughout the region, from lowlands into the subalpine zone.

Scheuchzer's Cotton-Grass
Eriophorum scheuchzeri

Narrow-Leaved Cotton-Grass
Eriophorum angustifolium

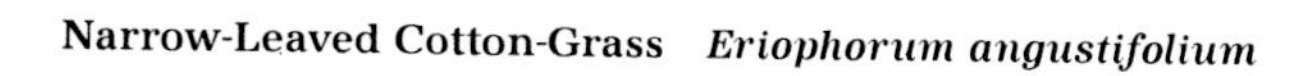

Narrow-Leaved Cotton-Grass *Eriophorum angustifolium*

Rush Family Juncaceae

The rushes show much variation in their overall appearance. Bullrushes, for example, are tall, coarse aquatics with round leaves that look like extensions of the stem—the inflorescence seems to be growing out of the side of the stem. And woodrushes look very much like grasses. In spite of these differences, floral characteristics in the family are consistent. The flowers are complete, with sepals, petals, stamens, and a pistil. However, the three sepals and three petals are reduced and bractlike, a characteristic associated with wind pollination. Most rushes are lowland aquatics and are not treated in this book, but some of them are important in various mountain habitats. For a comparison of the characteristics of rushes and other graminoids, see the table on page 359.

Juncaceae
(Rushes)

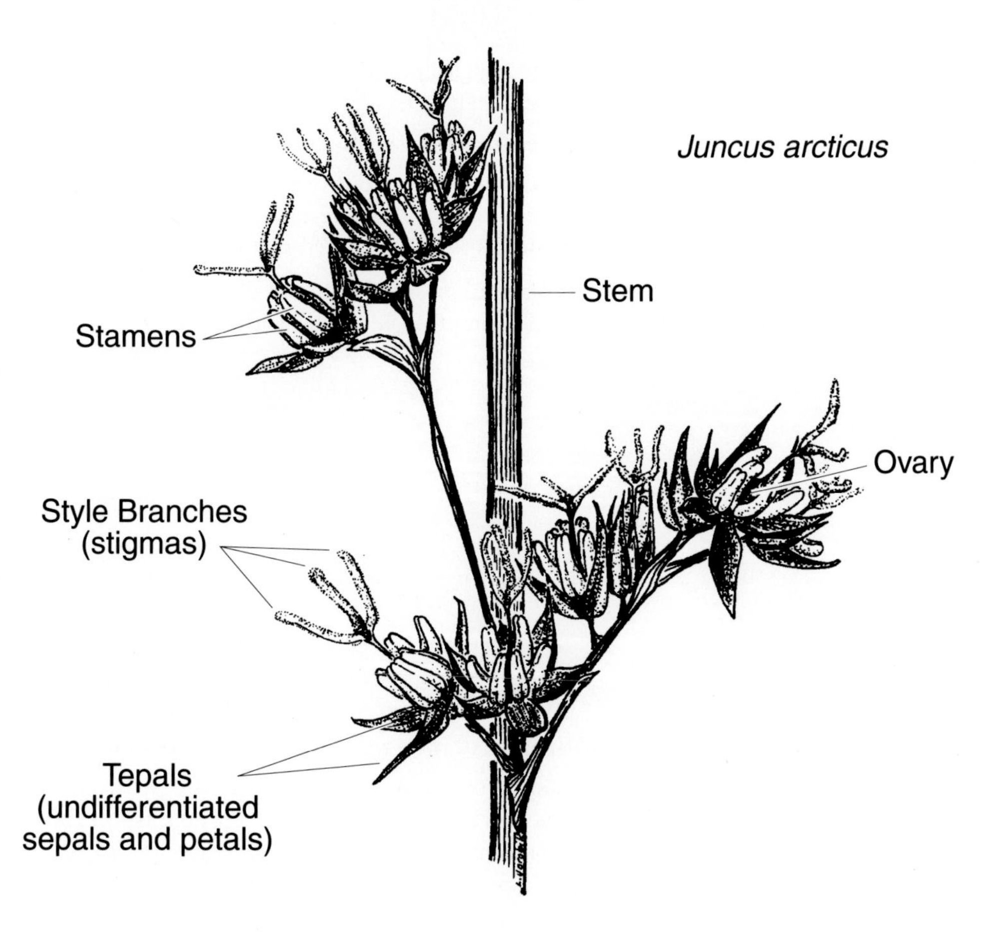

Small-Flowered Woodrush *Luzula parviflora*

Small-flowered woodrush has weak, tufted, spreading stems 8 to 32 inches (20 to 80 cm) long.

Leaves: Flat and lax with a few white hairs on the margins, about ⅜ inch (5 to 10 mm) wide, mainly basal but with four or more stem leaves.

Inflorescence: An open, nodding panicle with the flowers single or paired at the end of threadlike stalks; floral bracts (reduced sepals and petals) purplish brown and about as long as the ovary; frilly edged bracts are borne below the individual flowers.

Fruits: Dark brown, egg-shaped capsules, each with three seeds.

Ecology: Common at all elevations throughout the Pacific Northwest in moist forests, meadows, and along streamsides; particularly abundant on moist, north-facing slopes in areas of late snowmelt.

Piper's woodrush *(L. piperi)* is a similar species, but is shorter, with fewer (three), narrower stem leaves, and is mainly a high-elevation species.

Many-Flowered Woodrush *Luzula multiflora*

A tufted, perennial plant with erect stems 4 to 24 inches (10 to 60 cm) tall.

Leaves: Flat or slightly inrolled, about ¼ inch (2 to 6 mm) wide, with a few white hairs on the margins and a tuft of woolly hair at the base of the leaf, in the leaf axis; leaves mainly basal, the few stem leaves brownish, or at least with a brownish tip, and rather succulent.

Inflorescence: Flowers congested into a few to several heads; floral bracts (sepals/petals) pale green to chestnut brown, about ⅛ inch (2 to 4.5 mm) long, often with translucent margins; frilly edged bracts are borne below the individual flowers.

Fruits: Brown, egg-shaped capsules, each producing three seeds.

Ecology: Common throughout the Pacific Northwest in dry open forests, along rocky slopes, and in dry meadows in all vegetational zones.

Small-Flowered Woodrush
Luzula parviflora

Many-Flowered Woodrush
Luzula multiflora

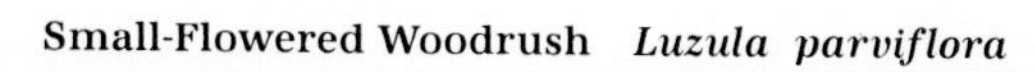

Small-Flowered Woodrush *Luzula parviflora*

Mertens' Rush *Juncus mertensianus*

This rush is a densely tufted plant, with numerous erect and spreading stems 2 to 16 inches (5 to 40 cm) long.

Leaves: More or less rounded, mainly basal; stem leaves few (one to four), the upper, bractlike stem leaf borne immediately below the inflorescence.

Inflorescence: A solitary, terminal, rounded, blackish brown head, with numerous congested flowers; blackish bractlike sepals/petals about ⅛ inch (3 to 4 mm) long.

Fruits: Oblong, egg-shaped, many-seeded capsules.

Ecology: Common throughout the Pacific Northwest in moist meadows and areas of late snowmelt, along stream banks, and around ponds, from mid-elevations to the alpine zone.

Although several other rushes in the region have rounded inflorescences, none has a solitary, terminal head. Many rushes have inflorescences that appear to be growing out of the side of the stem. This characteristic is exemplified by **arctic rush** *(J. arcticus),* a coarse, strongly rhizomatous plant of wet places from low to mid-elevations throughout the range (see illustration on page 389). The stems of arctic rush typically grow in a row, arising from the creeping rhizomes.

Parry's Rush *Juncus parryi*

The stems of this rush are strongly tufted and often cushionlike, 2 to 12 inches (5 to 30 cm) high.

Leaves: Mainly basal, reduced to sheaths with bristly tips about ⅜ inch (1 cm) long; stem leaves few, round, and up to 3 inches (8 cm) long.

Inflorescence: Appears to be lateral because of the extended, round, sharp-pointed leaf (called an involucral bract); one to three flowers per inflorescence; each flower has brownish bracts directly below its dark brown, bractlike sepals/petals.

Fruits: Oblong, egg shaped, many-seeded capsules.

Ecology: A common rush throughout southern regions of the Pacific Northwest, in dry meadows and on rocky slopes in the subalpine and alpine zones.

Drummond's rush *(J. drummondii)* is more widespread than Parry's rush and is easily mistaken for it. It has much shorter involucral bracts (rarely longer than the inflorescence), lacks definite leaf blades, and grows in moister habitats.

Mertens' Rush *Juncus mertensianus*

Arctic Rush *Juncus arcticus*

Parry's Rush *Juncus parryi*

How to Use the Keys

Although attempting to use an identification key can be frustrating for the novice botanist, familiarity with the key and with plant structures in general should resolve most difficulties. You will become familiar with the keys by using them. The illustrations that precede the glossary will acquaint you with plant structures.

The keys presented here are dichotomous, or two branched; that is, at every position in the key, the user has two mutually exclusive choices, "a" or "b" of the same number. To identify a plant, the user should start at the beginning of the appropriate key and always move forward (never backward). Choose between the two descriptions of each numerical set, note the number listed after your choice, and progress to the two choices marked by that number. Continue to make such choices until the choice you select refers to a family name or plant name and the page where the family or plant is described and pictured in the book. If only the family name is given in the key, refer to the descriptions and photographs of plants listed in the section on that family to identify the plant.

A few notes to help you use the keys: (1) read the descriptions of "a" and "b" carefully before making a choice; (2) always choose between "a" and "b" of the same number; (3) after choosing between "a" and "b" proceed to the numeral indicated until you have identified the plant or the plant's family; (4) qualifying words such as "mostly," "often," "usually," etc., mean what they say and should not be ignored.

Family descriptions are based on representatives from the mountains of the Pacific Northwest. If all species within a family were taken into consideration, the variation would be much greater and the families would be much harder to key out.

Key to Identifying Ferns and Fern Allies

1a. Fronds once pinnate (pinnae sometimes lobed) 2

1b. Fronds 2 or more times pinnate 4

2a. Fronds dimorphic; fertile (spore-bearing) fronds erect, with narrow pinnae; sterile fronds spreading, with broade pinnae .... **Deer fern (p. 36)**

2b. Fronds all alike and fertile 3

3a. Pinnae with a short stalk; indusia present .... **Sword fern (p. 42)**

3b. Pinnae nonstalked; indusia absent .... **Licorice fern (p. 42)**

4a. Blade of frond as wide as it is long, kidney shaped in outline; stipes black and shiny .... **Maidenhair fern (p. 34)**

4b. Blade of frond mostly longer than wide, elliptical or triangular in outline; stipes *usually* not black 5

5a. Fronds dimorphic; sterile frond parsleylike .... **Parsley fern (p. 36)**

5b. Fronds all alike, variable in shape 6

6a. Fronds borne singly, blades triangular in outline 7

6b. Fronds borne in clusters of two or more; blades *usually* not triangular 8

7a. Fronds less than 20 inches (0.5 m) tall; stipes yellowish brown .... **Oak fern (p. 40)**

7b. Fronds more than 20 inches (0.5 m) tall; stipes green to straw colored .... **Bracken fern (p. 44)**

8a. Stipes reddish brown to black, at least at the base; wiry and shiny, persisting after the blades die 9

8b. Stipes usually green and not persistent 10

9a. Lower surface of blades woolly .... **Lip-fern**

9b. Lower surface of blades not woolly .... **Woodsia (p. 38)**

10a. Frond outline usually triangular;
indusium horseshoe shaped .... **Wood fern (p. 40)**

10b. Frond outline elliptical; indusium flaplike 11

11a. Fronds small and delicate, 2 to 8 inches
(5 to 20 cm) long .... **Bladder-fern**

11b. Fronds larger, more than 8 inches (20 cm) long ...... **Lady-fern (p. 34)**

Key to Identifying Trees

1a. Plants usually evergreen; leaves needlelike or scalelike (conifers) 2

1b. Plants deciduous; leaves broad, not as above (hardwoods–angiosperms) 19

2a. Leaves scalelike; cones less than ½ inch (15 mm) long 3

2b. Leaves needlelike; cones more than ½ inch (15 mm) long 4

3a. Cones round; branches typically "weeping" .... **Alaska cedar (p. 48)**

3b. Cones ovate; branches spraylike .... **Western red cedar (p. 50)**

4a. Needles in clusters of 2 or more 5

4b. Needles borne singly, arranged spirally 9

5a. Needles several per cluster, deciduous in winter .... **Larches (p. 60)**

5b. Needles 2 to 5 per cluster, evergreen (pines) 6

6a. Needles 5 per cluster 7

6b. Needles 2 or 3 per cluster 8

7a. Cones elliptical, 6 inches (15 cm) long or longer; needles blue-green; trees of mid-montane habitats .... **Western white pine (p. 70)**

7b. Cones ovate, less than 6 inches (15 cm) long; needles green; plants of subalpine or alpine habitats .... **Whitebark pine (p. 66)**

8a. Needles 5 to 10 inches (13 to 25 cm) long, usually 3 per cluster; cones 3 to 6 inches (8 to 15 cm) long .... **Ponderosa pine (p. 72)**

8b. Needles less than 4 inches (10 cm) long, usually 2 per cluster; cones ¾ to 2 inches (2 to 5 cm) long .... **Lodgepole pine (p. 68)**

9a. Seeds with a bright red, pulpy covering (berrylike); needles with a petiole and abrupt, soft point .. **Western yew (p. 80)**

9b. Seeds borne in woody cones; needles not as above 10

10a. Cones with 3-pronged, conspicuous bracts; buds sharp pointed and chestnut colored .... **Douglas fir (p. 78)**

10b. Cones lacking conspicuous 3-pronged bracts; buds not both sharp pointed and chestnut colored 11

11a. Needles 4-sided, stiff, and more or less spiny; twigs roughened by persistent leaf bases (spruces) 12

11b. Needles 2-sided, more or less flexuous and blunt, not spiny; twigs mainly smooth (but see hemlocks) 14

12a. Needles somewhat flattened, whiter on the upper surface; cones usually more than 2 inches (5 cm) long .... **Sitka spruce (p. 62)**

12b. Needles quadrangular, equally white on all surfaces; cones usually less than 2 inches (5 cm) long 13

13a. Cone scales rounded and smooth; plants of northern British Columbia and Alaska .... **White spruce (p. 64)**

13b. Cone scales pointed, ragged along the tip; plants of southern British Columbia and Washington .... **Engelmann spruce (p. 62)**

14a. Cones pendant, borne throughout the tree; cone from persistent leaf bases (hemlocks) 15

14b. Cones standing erect in the tops of trees; cone scales deciduous, thus cones do not fall as a unit; twigs more or less smooth (true firs) 16

15a. Needles appearing 2-ranked (oriented on 2 sides of the branches), whitish on the lower surface; cones ¾ to 1 inch (1.5 to 2.5 cm) long .... **Western hemlock (p. 74)**

15b. Needles surrounding the twigs, whitish on both surfaces; cones more than 1 inch (2.5 cm) long .... **Mountain hemlock (p. 76)**

16a. Needles whitish below, shiny green above, appearing 2-ranked 17

16b. Needles whitish on both surfaces, not 2-ranked 18

17a. Needles concentrated on the top of the twig (hiding the twig) and projected forward .... **Pacific silver fir (p. 52)**

17b. Needles projected outward on the sides, the twig not hidden .... **Grand fir (p. 54)**

18a. Cones 4½ to 7 inches (11 to 18 cm) long; cone scales spine tipped; tree crown pyramidal; trees at mid-elevations in the southern Cascades of Washington .... **Noble fir (p. 56)**

18b. Cones 2½ to 4½ inches (6 to 11 cm) long; cone scales not spine tipped; tree crown very narrow and spirelike; trees in high montane and subalpine habitats throughout the range .... **Subalpine fir (p. 58)**

19a. Leaves opposite, palmately lobed or divided (maples) 20

19b. Leaves alternate, entire, toothed or pinnately divided 22

20a. Leaves with 5 to 7 major lobes; flowers reddish .... **Vine maple (p. 88)**

20b. Leaves with 3 major lobes; flowers greenish yellow 21

21a. Leaves often more than 6 inches (15 cm) broad; flowers in dense, pendant clusters .... **Big-leaf maple (p. 90)**

21b. Leaves less than 6 inches (15 cm) broad; flowers few, in nonhanging clusters .... **Douglas maple (p. 88)**

22a. Fruits fleshy; flowers often showy, white to greenish 23

22b. Fruits not fleshy; flowers nonshowy, lacking petals, and borne in catkins 24

23a. Flowers greenish; leaves nearly opposite .... **Cascara (p. 96)**

23b. Flowers white; leaves clearly alternate .... **Bitter cherry (p. 268)**

24a. Leaves conspicuously toothed; plants bisexual, male and female catkins borne on the same plant 25

24b. Leaves entire or inconspicuously toothed; plants unisexual, male and female catkins borne on separate plants 26

25a. Teeth of leaves *usually* rounded; female catkins woody and conelike, persistent on the trees .... **Alders (p. 84)**

25b. Teeth of leaves sharp pointed; female catkins not woody nor persistent on trees .... **Birches (p. 86)**

26a. Leaf blades twice as long as broad; plants shrublike, with multiple stems .... **Willows (p. 92)**

26b. Leaf blades about as long as broad; plants with a single stem (trunk) 27

27a. Leaf petioles flattened; tree trunk smooth, pale green to white .... **Quaking aspen (p. 92)**

27b. Leaf petioles rounded; tree trunk fissured, gray to gray-green .... **Black cottonwood (p. 94)**

Key to Identifying Forbs and Shrubs

1a. Medium-sized to large shrubs, 2½ feet (75 cm) tall or more, usually in or bordering forests or occasionally in mountain meadows 2

1b. Smaller shrubs *or* herbs, in various habitats including forests and meadows 10

2a. Flowers individually inconspicuous, unisexual, and borne in catkins; leaves alternate 3

2b. Flowers mostly showy, usually bisexual, and borne singly or variously clustered; leaves alternate *or* opposite 4

3a. Fruit a capsule with numerous cottony seeds; plants not particularly aromatic (willows) .... **Willow family (p. 348)**

3b. Ovary (fruit) 1-seeded, fleshy when mature; plants aromatic (sweet gale) .... **Sweet Gale family (p. 336)**

4a. Stamens more than 10; petals not fused to each other; flowers large and showy with fleshy fruit *or* flowers small and densely clustered with nonfleshy fruit .... **Rose family (p. 262)**

4b. Stamens 10 or fewer; petals sometimes fused at the base; flower size and fruit type various 5

5a. Leaves opposite; fruit a berry 6

5b. Leaves alternate; fruit a capsule *or* berry 7

6a. Petals absent; leaves brown-scaly on the lower surface, simple (buffalo berry) .... **Oleaster family (p. 230)**

6b. Petals present, fused; leaves not scaly beneath, simple (twinberry) or pinnately compound (elderberry) .... **Honeysuckle family (p. 198)**

7a. Leaves pinnately compound, leathery, the leaflet margins spiny; flowers yellow (tall Oregon grape) .... **Barberry family (p. 104)**

7b. Leaves not as described above; flowers not yellow 8

8a. Leaves large, more than 8 inches (20 cm) wide; stems and leaf veins spiny (devil's club) ..... **Ginseng family (p.196)**

8b. Leaves less than 5 inches (12 cm) wide; stems *sometimes* spiny 9

9a. Leaves palmately lobed or divided (maplelike); fruit a berry; plants aromatic .... **Currant family (p. 134)**

9b. Leaves entire or shallowly toothed; fruit a berry *or* capsule; plants not aromatic .... **Heath family (p. 166)**

10a. Aquatic plants with heart-shaped, floating leaves and large, yellow flowers (pond lily) .... **Water Lily family (p. 346)**

10b. Plants not as described above 11

11a Flowers densely clustered into a spike, with an associated yellow modified leaf (spathe); leaves large, up to 3 feet (1 m) long and half as wide (skunk cabbage) .... **Arum family (p. 102)**

11b. Flowers not associated with a yellow spathe; leaves much smaller 12

12a. Plants white, yellow, or reddish, without green leaves or stems (nonphotosynthetic) 13

12b. Plants with green leaves and/or stems (photosynthetic) 14

13a. Flowers bilaterally symmetrical, with 3 petals and 3 sepals (coral root orchids) .... **Orchid family (p. 222)**

13b. Flowers radially symmetrical, with 5 petals and 5 sepals (nongreen ericads) .... **Heath family (p. 166)**

14a. Flowers densely clustered into heads that resemble single flowers having many petals and sepals (the "sepals" being green or papery involucral bracts); in general, the heads are daisylike, dandelion-like, or thistlelike; heads sometimes small and also clustered .... **Sunflower family (p. 302)**

14b. Flowers individually distinct and not clustered into heads *or if* densely clustered, not daisylike, dandelion-like or thistlelike 15

15a. Dwarf shrubs with inconspicuous flowers borne in catkins; leaves alternate, at least half as wide as long (willows) .... **Willow family (p. 348)**

15b. Plants differing from those described above in one or more characteristics 16

16a. Evergreen shrub with leathery, toothed leaves; flowers small, with 4 purplish petals (mountain box) .... **Mountain Box family (p. 216)**

16b. Plants not as described above in all respects 17

17a. Shrubs with large, evergreen, pinnately compound leaves; leaflets spiny along the margin; flowers yellow, with 9 or more petals (low Oregon grape) .... **Barberry family (p. 104)**

18a. Shrubs with scalelike or needlelike leaves; flowers bell shaped or small and inconspicuous 19

18b. Herbs, or, if shrubby, leaves other than described above *or* flowers not bell shaped 20

19a. Flowers bell shaped and showy; fruit a capsule (bog laurel and heathers) .... **Heath family (p. 166)**

19b. Flowers inconspicuous, purplish; fruit a black berry (crowberry) .... **Crowberry family (p. 196)**

20a. Shrubs with ovate or elliptical leaves and urn-shaped or bell-shaped flowers; fruit a berry .... **Heath family (p. 166)**

20b. Plants not as described above in all characteristics 21

21a. Dwarf shrub with whorled leaves; flowers small, bordered by 4 large, white, petal-like bracts; fruit a red-orange berry (ground dogwood).... **Dogwood family (p. 138)**

21b. Plants not as described above 22

22a. Flower parts in multiples of 3 (4 in false lily-of-the-valley)–3 petals and 3 sepals, usually 3 or 6 stamens; leaves usually with parallel venation, simple; herbs 23

22b. Flower parts in multiples of 4 or 5 *and/or* leaves with net venation, often compound; herbs or shrubs 24

23a. Flowers bilaterally symmetrical; ovary inferior; fruit a capsule .... **Orchid family (p. 222)**

23b. Flowers radially symmetrical; ovary superior; fruit a capsule or berry .... **Lily family (p. 202)**

24a. Flowers with 3 purplish, thread-tipped sepals (petals absent); leaves heart shaped (wild ginger) .... **Birthwort family (p. 102)**

24b. Plants not as described above 25

25a. Flowers bilaterally symmetrical, individually distinct 26

25b. Flowers radially symmetrical, *often* small and densely clustered 32

26a. Petals not fused; leaves simple, ovate or heart shaped; stamens 5 (violets) .... **Violet family (p. 338)**

26b. Petals fused, at least in part, *or* leaves compound *and* stamens 10 or more 27

27a. Stamens more than 10; sepals more colorful than petals (larkspurs and monkshoods) .... **Buttercup family (p. 116)**

27b. Stamens 10 or fewer; petals more colorful than sepals 28

28a. Leaves divided into numerous small segments, fernlike, basal; sepals 2 (bleedingheart) ... **Fumitory family (p. 138)**

28b. Leaves not as above, though often pinnately or palmately compound; sepals 4 or 5, though often fused and the number obscure, or modified into bristles 29

29a. Stamens 10; some petals free (not fused) or at least the petals not readily separating from the flower as a unit .... **Pea family (p. 234)**

29b. Stamens 2, 4, or 5; petals all fused, separating from the flower as a unit 30

30a. Ovary inferior; sepals modified into numerous bristles (Sitka valerian)................................ **Valerian family (p. 336)**

30b. Ovary superior; sepals not modified into bristles 31

31a. Flowers borne singly on leafless stems, blue; leaves all basal and insectivorous (butterwort)... **Bladderwort family (p. 106)**

31b. Flowers not borne singly on leafless stems *or* color not blue; plants not insectivorous ..... **Figwort family (p. 144)**

32a. Flowers borne in large umbels, 6 inches (15 cm) or more across; leaves 10 inches (25 cm) or more wide, palmately divided (cow parsnip) .... **Parsley family (p. 230)**

32b. Plants not as above 33

33a. Flowers very small, reddish, with 4 sepals and no petals; leaves basal, broadly heart shaped (mountain sorrel)..................................... **Buckwheat family (p. 112)**

33b. Plants not as described above in all characteristics........................... 34

34a. Flowers borne in umbels; leaves compound, parsleylike; ovary inferior .... **Parsley family (p. 230)**

34b. Flowers not borne in umbels, *or* if so, leaves simple and ovary superior 35

35a. Flowers small, greenish or white (to pink), densely clustered in a round to elongate inflorescence, the individual flowers barely distinct 36

35b. Flowers larger (color varied) and individually distinct *or* inflorescence more open (but see saxifrage family) 40

36a. Leaves simple and not lobed; plants *usually* mat forming 37

36b. Leaves compound or palmately lobed; plants not mat forming 38

37a. Sepals 2, papery in texture, white to pink, larger than the 5 petals; prostrate herbs (pussypaws) .... **Purslane family (p. 258)**

37b. Sepals absent; plants erect herbs with 5 petals (mountain bistort) *or* plants dwarf shrubs with 6 petals in two whorls (desert buckwheats) .... **Buckwheat family (p. 112)**

38a. Leaves divided into 3 large, wedge-shaped leaflets; inflorescence and leaf blades borne on separate long stalks from the root crown (vanilla leaf) .... **Barberry family (p. 104)**

38b. Leaves pinnately compound or palmately divided; stems leafy 39

39a. Leaves strictly pinnately compound; flowers greenish to pink (Sitka burnet) .... **Rose family (p. 262)**

39b. Leaves divided into multiples of 3 segments (baneberry) or palmately divided (false bugbane); flowers white .... **Buttercup family (p. 116)**

40a. Petals and sepals 4; stamens in even numbers 41

40b. Petals and sepals usually 5 (not 4, petals sometimes absent); stamens usually in odd numbers 44

41a. Leaves whorled; ovary inferior; stamens 4 (northern bedstraw) .... **Madder family (p. 216)**

41b. Leaves alternate or opposite; ovary superior *or* inferior; stamens 6 or 8 42

42a. Flowers blue, slightly irregular; stamens 2 (speedwells) .... **Figwort family (p. 144)**

42b. Flowers white, yellow or reddish; stamens 6 or 8 43

43a. Ovary inferior; stamens 8 .... **Evening-Primrose family (p. 140)**

43b. Ovary superior; stamens 6 .... **Mustard family (p. 218)**

44a. Petals absent, sepals green; plants unisexual, the flowers either bearing pendant, chandelier-like stamens or erect, reddish ovaries; leaves divided into multiples of 3 segments (western meadowrue) .... **Buttercup family (p. 116)**

44b. Plants not as described above in all characteristics 45

45a. Petals absent, sepals fused into a vaselike tube with 5 spreading lobes and 5 stamens borne on the rim of the tube; fruit a red berry (geocaulon) .... **Sandalwood family (p. 284)**

45b. Petals present and plants otherwise not as described above in all characteristics 46

46a. Petals fused, separating from the flower as a unit; stamens 5 (rarely 6) 47

46b. Petals free *or* if slightly fused at the base (ericads), stamens 10....... 57

47a. Ovary inferior; flowers bell shaped *and* nodding, blue or pale lavender, 1 or few per stem 48

47b. Ovary superior; flowers not bell shaped, nodding and blue *or* if so (bluebells), flowers several per stem 49

48a. Herbs with blue flowers (bellflower) .... **Harebell family (p. 164)**

48b. Trailing evergreen shrubs with pale lavender flowers, 2 per stem (twinflower)........ **Honeysuckle family (p. 198)**

49a. Leaves pinnately compound.. 50

49b. Leaves simple though sometimes toothed 51

50a. Stamens extended well beyond the petal lobes; styles 2 .... **Waterleaf family (p. 342)**

50b. Stamens not extended beyond the petal lobes; styles 3 (Jacob's ladder and sky pilot) .... **Phlox family (p. 242)**

51a. Flowers blue; flowering stems leafy 52

51b. Flowers white to red or lavender; flowering stems *often* naked (without leaves) 53

52a. Flowers bell shaped but erect; ovary ovate, maturing into a many-seeded capsule (gentians)........ **Gentian family (p. 160)**

52b. Flowers wheel shaped (the petal lobes perpendicular to the petal tube) *or* if bell shaped, then nodding; ovary 4-lobed, maturing into 4 nutlets .... **Borage family (p. 108)**

53a. Leaves whorled at the top of 6-inch (15 cm) stems (±); flowers borne singly on threadlike petioles from the stem tip; petals usually 6 (starflowers) .... **Primrose family (p. 254)**

53b. Plants not as described above 54

54a. Stems leafy; leaves linear, needlelike; styles 3 (phloxes) .... **Phlox family (p. 242)**

54b. Stems naked; leaves often broader; styles 1 or 2 55

55a. Plant a very small, branched annual with nonshowy white to reddish flowers (fairy-candelabra) .... **Primrose family (p. 254)**

55b. Plants perennial and flowers showy 56

56a. Flowers white; leaves broadly heart shaped to round (deer-cabbage) .... **Buckbean family (p. 106)**

56b. Flowers pink to red, sometimes facing downward with the petals swept backward (shootingstars); leaves much longer than broad .... **Primrose family (p. 254)**

57a. Stamens numerous, more than 10; ovaries 3 to many 58

57b. Stamens 10 or fewer; ovary usually 1, *sometimes* 2 or 5 59

58a. Sepals, petals, and stamens borne on the receptacle with no fusion of floral parts .... **Buttercup family (p. 116)**

58b. Sepals, petals, and stamens borne on a cup-shaped or saucer-shaped "calyx tube" that surrounds the multiple ovaries .... **Rose family (p. 262)**

59a. Flowers yellow; ovaries 5; plants very succulent (stonecrops) .... **Stonecrop family (p. 298)**

59b. Flowers rarely yellow; ovaries 1 or 2; plants *sometimes* succulent 60

60a. Leaves palmately lobed or divided (± maplelike) or trifoliate (cloverlike) 61

60b. Leaves sometimes toothed but not lobed or divided 63

61a. Flowers pink, about 1 inch (2.5 cm) across (wild geraniums) .... **Geranium family (p. 162)**

61b. Flowers usually white (not pink), much smaller 62

62a. Dwarf, matted plants with cloverlike leaves (sibbaldia) .... **Rose family (p. 262)**

62b. Plants sometimes mat forming but leaves not cloverlike, although sometimes 3 lobed .. **Saxifrage family (p. 286)**

63a. Flowers pink to dark red, or white with pink stripes 64

63b. Flowers white to greenish, rarely yellow 69

64a. Petals 6 to 10; leaves elongate (± linear), succulent, mostly basal; sepals 2 (lewisias) .... **Purslane family (p. 258)**

64b. Petals 5; leaves not both linear and succulent, though sometimes basal; sepals 2 or 5 65

65a. Low, very dense cushion plants; leaves less than ½ inch (12 mm) long 66

65b. Plants mostly erect; leaves longer than ½ inch (12 mm) 67

66a. Petals 2 lobed; leaves linear to lance shaped; sepals fused (moss campion) .... **Pink family (p. 246)**

66b. Petals not lobed; leaves ovate; sepals not fused (purple saxifrage) .... **Saxifrage family (p. 286)**

67a. Flowers nodding; sepals 5; ovary and style 1; plants not succulent (wintergreens and prince's pine) .... **Heath family (p. 166)**

67b. Flowers erect; sepals 2 or 5; ovaries 5 *or* styles 3; plants succulent 68

68a. Flowers dark red to purple; sepals 5; ovaries 5 (rose root) .... **Stonecrop family (p. 298)**

68b. Flowers white with pink stripes; sepals 2; ovary 1 .... **Purslane family (p. 258)**

69a. Sepals 2; stamens 5; styles 3 .... **Purslane family (p. 258)**

69b. Sepals 5; stamens usually 10; styles 1, 3, or 5 70

70a. Leaves opposite; styles 3 or 5; petals often lobed (thus heart shaped) .... **Pink family (p. 246)**

70b. Leaves alternate, basal or whorled; styles 1 or 2; petals not lobed 71

71a. Flowers nodding; style 1; leaves rarely toothed (woodnymph and wintergreens) .... **Heath family (p. 166)**

71b. Flowers erect; styles 2; leaves usually toothed .... **Saxifrage family (p. 286)**

Flowers

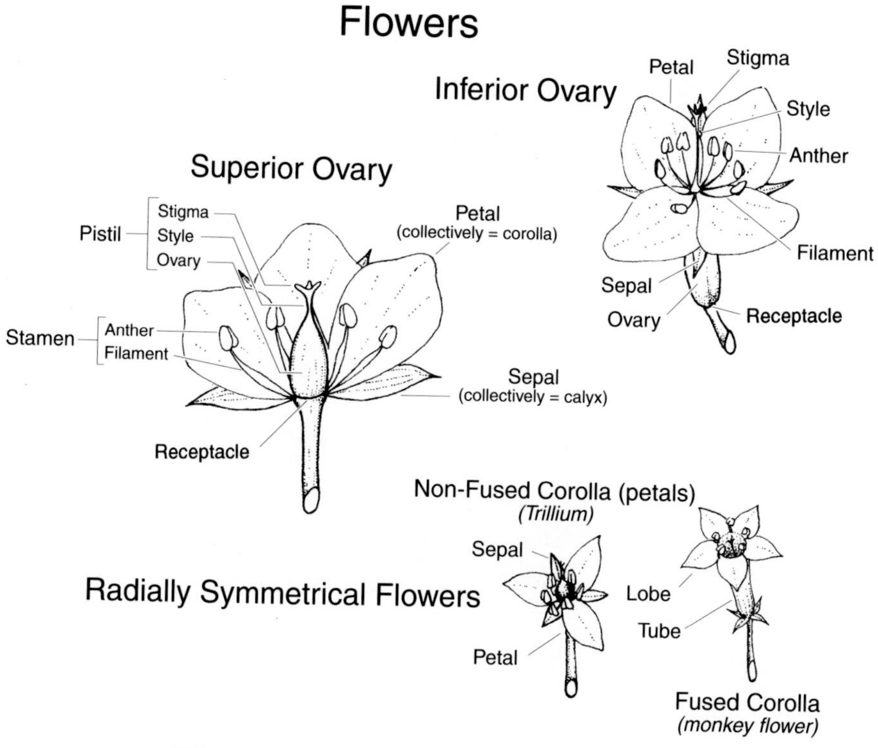

Bilaterally Symmetrical Flowers

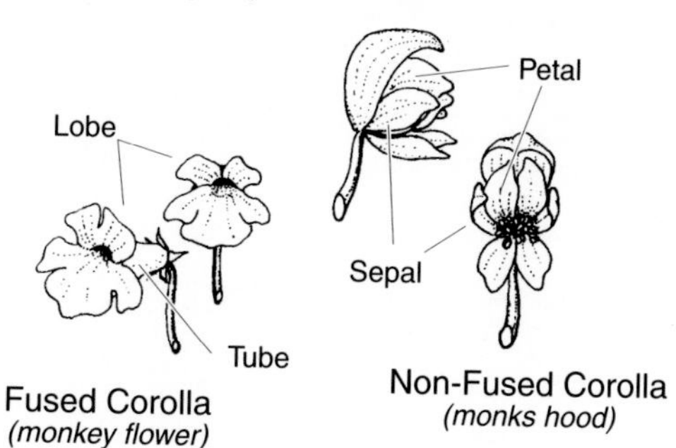

Inflorescence Types

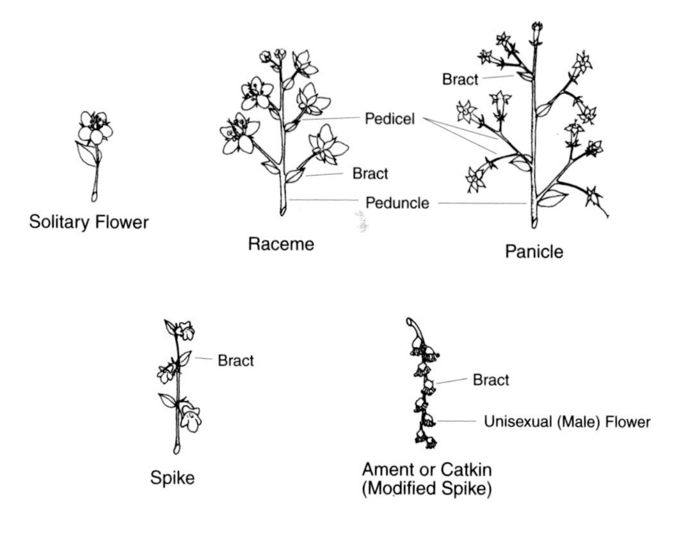

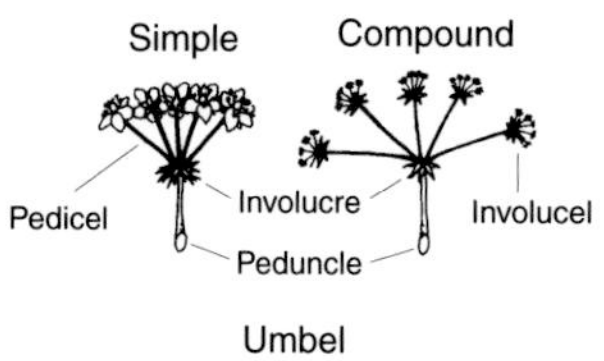

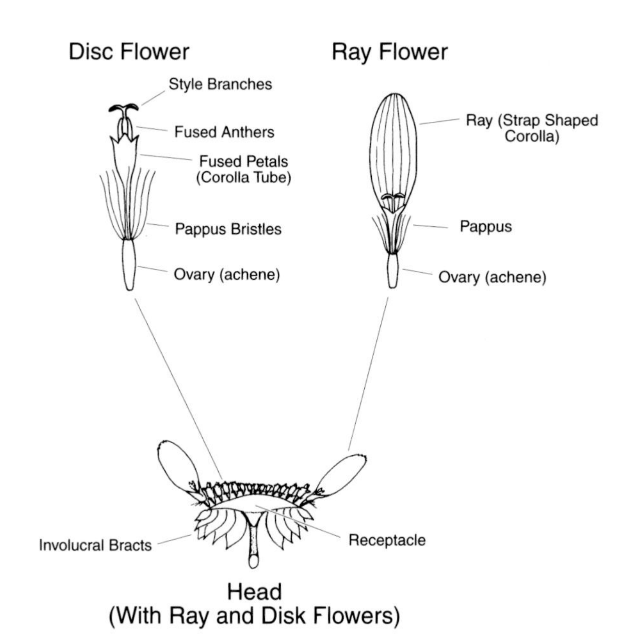

Leaves

Leaf Arrangement

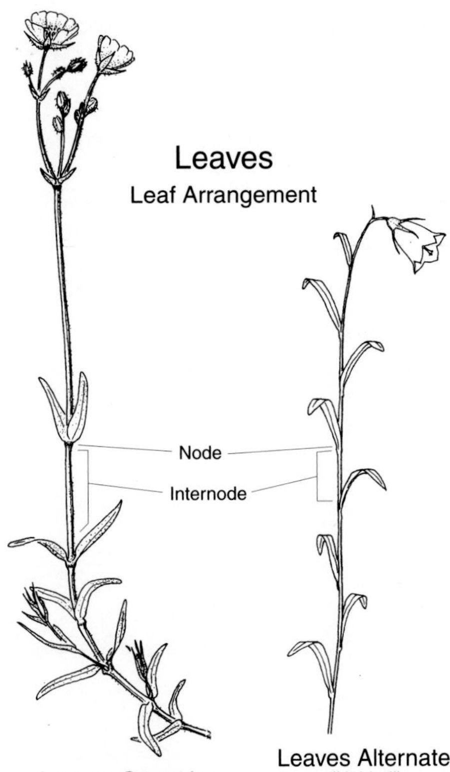

Leaves Opposite
(Field Chickweed)

Leaves Alternate
(Hairbell)

Leaves Whorled
(Starflower)

Leaves Basal
(Shooting Star)

Leaf Type/Shape

Simple Leaves

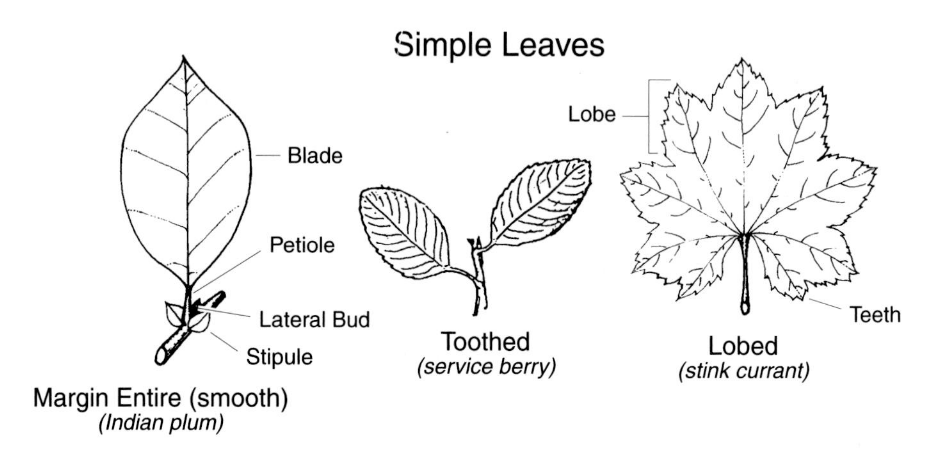

Compound Leaves

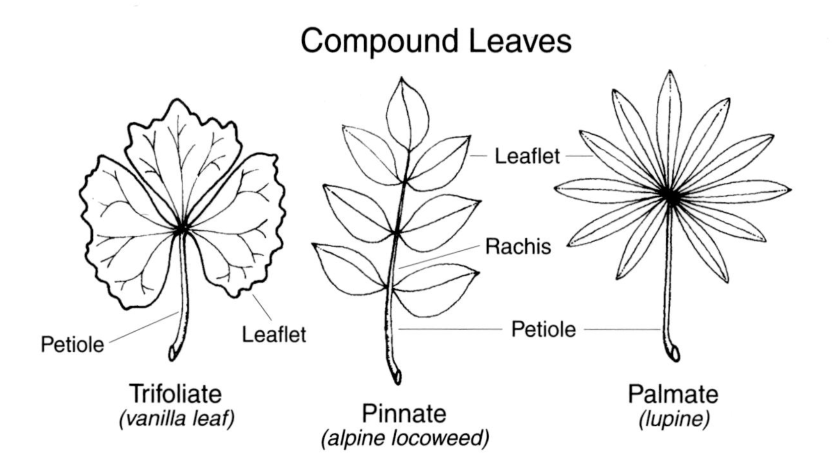

Venation

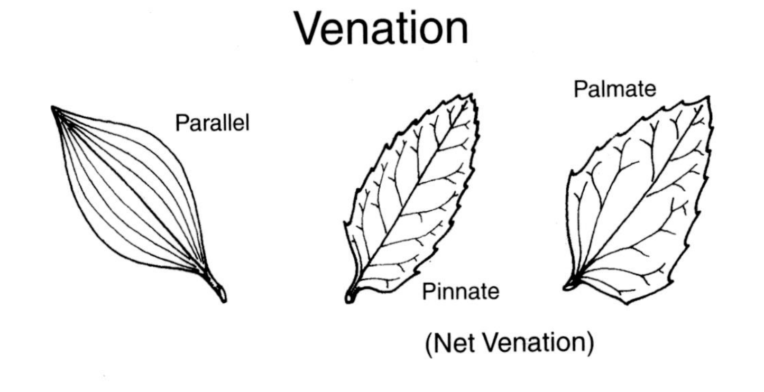

(Net Venation)

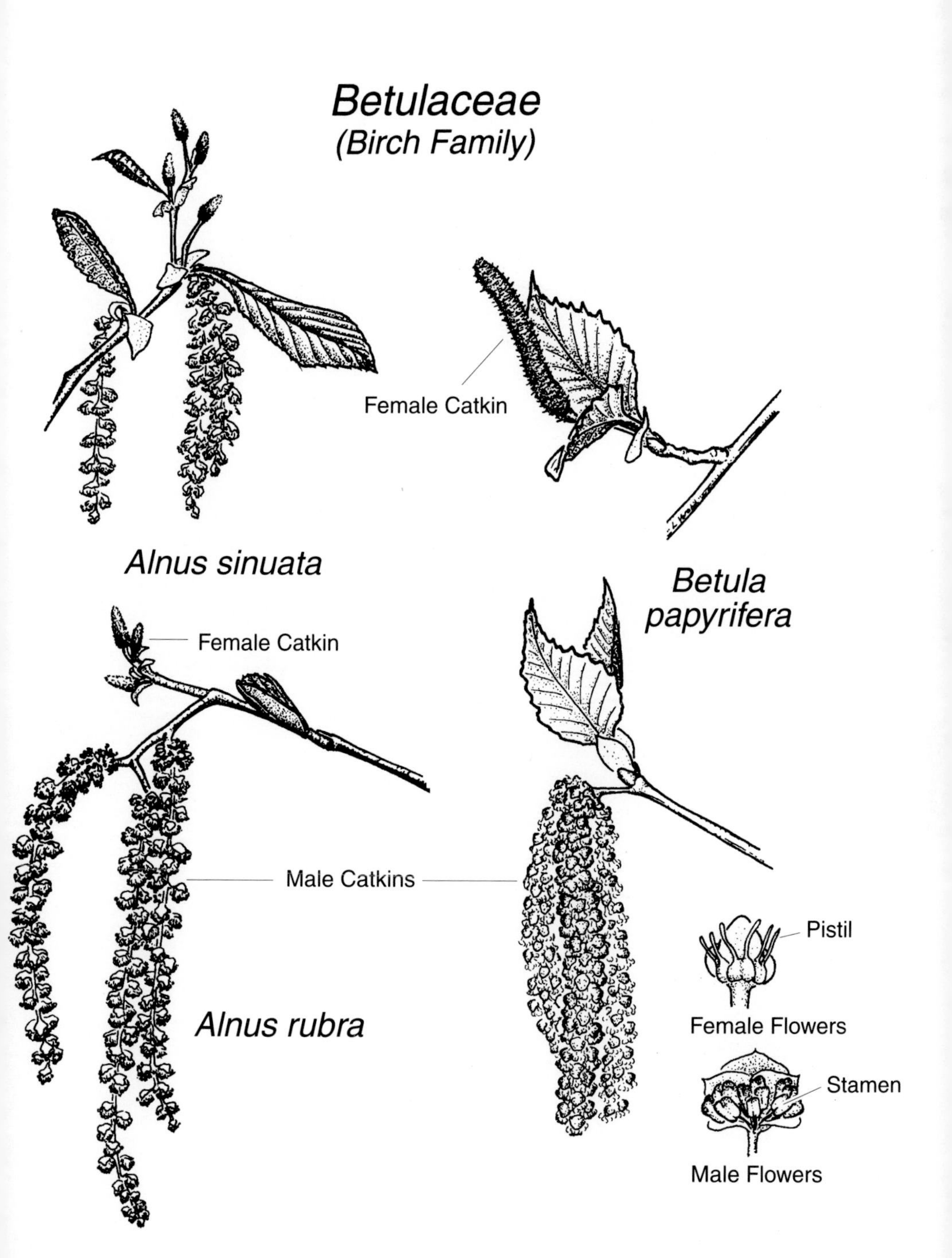
Betulaceae
(Birch Family)
Female Catkin
Alnus sinuata
Betula
papyrifera
Female Catkin
Male Catkins
Pistil
Alnus rubra
Female Flowers
Stamen
Male Flowers

Glossary

For explanations of technical terms used in the Glossary, see drawings in each plant section in the book.

achene A small, dry, one-seeded fruit that does not open at maturity.

alkaloid A toxic nitrogen-containing substance. Many plants produce alkaloids that are poisonous to animals—a defense mechanism.

alternate As applied to leaves, not opposite, one leaf per node.

angiosperm A flowering plant.

annual A plant that lives only one year (from seed to seed in one year).

apical At the apex or tip.

appressed Lying close or flat against a surface, such as hairs on a leaf.

awn A bristlelike appendage, usually at the apex of a structure.

axil The angle formed by a leaf with the stem.

basal As applied to leaves, at the base of the stem; at or near ground level.

biennial A plant that lives two years, the first year producing leaves and a thick taproot, the second year developing an erect stem with flowers.

bilaterally symmetrical Relating to a flower; irregular, with a left side and right side; mirror images can be produced only by dividing the flower in a vertical plane; one (or more) of the petals or sepals is unlike the others.

bilobed Divided into two lobes.

blade The flat part of a leaf or petal.

bract A small, modified, often pigmented leaf usually occurring at the base of a flower or flower cluster.

bulb A thickened, fleshy structure that is usually below ground and functions in food storage and reproduction.

capsule A fruit that becomes dry and splits open at maturity, shedding its seeds. It usually contains two or more compartments.

carpel Seed-bearing unit of a flower. A pistil is composed of one or more carpels.

climax species Species that perpetuate themselves indefinitely in a forest rather than being replaced by other species; as opposed to seral species.

community A group of plants living together in a given habitat.

conifer A cone-bearing plant, usually evergreen.

corolla A term for flower petals treated collectively, used mainly when the petals are fused.

cushion A growth form of some plants; dense and low in stature, resembling a cushion.

deciduous Falling away, as when leaves fall off at the end of the growing season.

disk flower One of the central flowers on the head of a sunflower, daisy, or similar plant; a tubular-shaped flower lacking a flattened extension (ray).

dissemination The act of spreading or scattering, such as seeds or pollen by animals or the wind.

dominant One of the most important plants in a given community because of its numbers or size, or both; a plant that has a major effect on other plants in the community.

elliptical Longer than wide with similar ends (not egg shaped); a squashed circle.

elongate Longer than wide.

entire With a smooth margin, neither toothed nor lobed.

fleshy Thick and succulent; containing juices (water).

flexuous Limber, easily bent.

floret Tiny flowers, used mainly to denote the flowers and associated bracts of a grass.

forb Herb, excluding ferns and grasslike plants.

gametophyte The sexual, gamete-bearing generation of a plant; applicable to the life cycle of ferns and fern allies.

glabrous Lacking hairs.

glandular Having glands that secrete resinous, often sticky material.

glume One of the two sterile bracts at the base of a spikelet in grasses.

grasping To appear to grasp the stem, as with some leaf blades that partly or totally encircle the stem.

gymnosperm A nonflowering seed plant; conifers are gymnosperms.

habit Growth form of a plant.

habitat The home of a given plant, unique in having a particular set of environmental conditions.

head A dense cluster of flowers that lack stalks; the inflorescence of a member of the sunflower family.

herb A plant lacking a hard, woody stem.

indusium An outgrowth covering and protecting the spore cluster in ferns.

inferior As related to an ovary, the flower parts borne on (above) the ovary or, conversely, the ovary borne below (and inferior to) the flower parts.

inflorescence A flower cluster.

irregular Pertaining to a flower, bilaterally symmetrical.

krummholz The low, spreading growth form of some conifers at or above the treeline.

leaflet One of the leaflike segments of a compound leaf.

lemma The lower and usually larger of the two bracts that immediately encloses the individual grass flower.

ligule In grasses, the flat, usually membranous projection from the summit of the sheath; in the sunflower family, the strap-shaped corolla of the ray flower.

linear Long and very narrow.

lobed As applied to a leaf, cut or dissected (but not all the way to the midvein of the leaf).

mat-forming Low, dense, and spreading horizontally; resembling a mat or carpet.

montane On or of the mountains, between lowlands and the subalpine zone.

nectary A specialized structure of a flower that produces nectar.

nodding As related to a flower, hanging with the face of the flower downward.

nutlet Small, hard-shelled, one-seeded fruit, as in the borage family.

oblong Longer than wide.

opposite As related to leaves, paired at the nodes; two leaves per node.

ovary The seed-containing part of the flower; the part of the flower that matures into a fruit.

ovate Egg shaped.

palmate Shaped like the palm of a hand with extended fingers.

palmately compound As applied to leaves, divided to the midvein in such a way that the leaflets are borne at the same point and spread like fingers.

parasite Growing on and deriving nourishment from another living plant.

perigynium (plural, *perigynia*) The sac, usually inflated, enclosing the ovary in sedges.

pedicel The stalk of a flower or fruit.

perennial A plant that lives more than two years; it may die down to the roots each year but sprouts up the next.

petal One of the bractlike segments of the inner whorl of flower parts, usually colored or showy.

petiole Leaf stalk.

pinnate Having two rows of lateral appendages along an axis, like a feather.

pinnately compound As applied to leaves; divided to the midvein with the leaflets arranged on both sides of the extended axis of the petiole.

pistil The central (female) part of the flower consisting of the ovary, style, and stigma.

pubescence General term for hairiness or woolliness.

pubescent Bearing hairs of any sort; hairy.

raceme An elongate, unbranched flower cluster, each flower having a stalk or pedicel.

ray The bladelike extension of a ray flower.

ray flower One of the outer flowers of a sunflower, daisy, or similar plant, which has a flattened, elongate, colorful extension.

reflexed Abruptly bent or turned backward or downward.

regular As related to a flower, radially symmetrical.

rhizome An underground, spreading stem that produces roots and upright branches.

root crown The juncture between the root and stem; the crown of the root.

rootstalk A rhizome.

runner A prostrate stem that roots at the nodes and produces erect branches (as with strawberries).

sepal One of the bractlike segments of the outer whorl of flower parts, usually green.

seral Of a stage, or sere, preceding and replaced by the climax stage in an ecosystem; also relating to a tree species, such as red alder, that ultimately is replaced in the development of a forest.

shrub A woody plant that branches at or near ground level.

sorus (plural, *sori*) The location on a fern frond where the spore sacs (sporangia) are clustered.

spike An elongate flower cluster with sessile (nonstalked) flowers.

sporangia Spore sacs.

spur A hollow extension of a petal or sepal, often containing nectar.

stamen The pollen-containing part of a flower.

steppe A nonforested region dominated by grasses and low shrubs.

stigma The pollen-receptive part of a pistil.

stipule Leaflike or bractlike appendage at the base of the petioles of some leaves. They occur in pairs.

stolon Horizontal, spreading stem that roots at the nodes; a runner.

style The narrow portion of the pistil, connecting the ovary with the pollen-receptive stigma.

succulent Soft and juicy; filled with water.

superior As applied to an ovary, the flower parts borne on the receptacle below the ovary; conversely, the ovary above (and superior to) the other flower parts.

talus Loose gravel or boulders on a slope.

taproot An elongate, unbranched, vertical root, like a carrot.

umbel An umbrella-shaped flower cluster or inflorescence.

unisexual One sex as a staminate (male) or pistillate (female) flower or plant.

whorl A group of three or more leaves, flowers, or petals radiating from a single point such as a node.

Recommended Reading

Arno, S. F. *Northwest Trees*. Seattle, Wash.: The Mountaineers, 1977.

Clark, L. J. *Wildflowers of British Columbia*. Sidney, B.C.: Gray's Publishing, 1973.

Douglas, G. W., G. B. Straley, and D. Meidinger. *The Vascular Plants of British Columbia*, Parts 1–4. Victoria, B.C.: British Columbia Ministry of Forests, 1989–1994.

Hitchcock, C. L., and A. Cronquist. *Flora of the Pacific Northwest*. Seattle, Wash.: University of Washington Press, 1973.

Hitchcock, C. L., A. Cronquist, M. Ownbey, and J. W. Thompson. *Vascular Plants of the Pacific Northwest*. Vol. 1-5. Seattle, Wash.: University of Washington Press, 1955–1969.

Horn, E. L. *Coastal Wildflowers of the Pacific Northwest*. Missoula, Mont.: Mountain Press, 1993.

Hulten, E. *Flora of Alaska and Neighboring Territories*. Stanford, Calif.: Stanford University Press, 1968.

Kozloff, E. N. *Plants and Animals of the Pacific Northwest, an Illustrated Guide to the Natural History of Western Oregon, Washington, and British Columbia*. Seattle, Wash.: University of Washington Press, 1976.

Larrison, Earl, Jr. *Washington Wildflowers*. The Trailside Series. Seattle, Wash.: Seattle Audubon Society, 1974.

Lellinger, D. B. *A Field Manual of the Ferns and Fern-Allies of the United States and Canada*. Washington, D.C.: Smithsonian Institution Press, 1985.

Mackinnon, A., J. Pojar, and R. Coupe. *Plants of Northern British Columbia*. Vancouver, B.C.: Lone Pine Publishing, 1992.

Mathews, D. *Cascade-Olympic Natural History*. Portland, Ore.: Raven Editors and Portland Audubon Society, 1988.

Pojar, J., and A. Mackinnon. *Plants of Coastal British Columbia*. Vancouver, B.C.: Lone Pine Publishing, 1994.

Taylor, R. J. *Northwest Weeds, The Ugly and Beautiful Villains of Fields, Gardens, and Roadsides*. Missoula, Mont.: Mountain Press, 1990.

——. *Sagebrush Country, A Wildflower Sanctuary*. Missoula, Mont.: Mountain Press, 1992.

Taylor, T. M. C. *The Lily Family (Liliaceae) of British Columbia*. Handbook No. 19. Victoria, B.C.: Royal British Columbia Museum, 1966.

——. *Ferns and Fern Allies of British Columbia*. Handbook No. 12. Victoria, B.C.: Royal British Columbia Museum, 1973.

——. *The Rose Family (Rosaceae) of British Columbia*. Handbook No. 30. Victoria, B.C.: Royal British Columbia Museum, 1973.

——. *The Pea Family (Leguminosae) of British Columbia*. Handbook No. 32. Victoria, B.C.: Royal British Columbia Museum, 1974.

——. *The Figwort Family (Scrophulariaceae) of British Columbia*. Handbook No. 33. Victoria, B.C.: Royal British Columbia Museum, 1974.

——. *The Sedge Family (Cyperaceae) of British Columbia*. Handbook No. 43. Victoria, B.C.: Royal British Columbia Museum, 1983.

Index

Taxonomic names in parentheses are "old" names, synonyms.
Boldface numbers signify pages with photos.

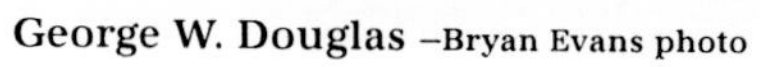

George W. Douglas –Bryan Evans photo

Ronald J. Taylor

About the Authors

Ronald J. Taylor, a Ph.D. professor of botany and plant ecology at Western Washington University, has lived and worked in the Pacific Northwest for over thirty years. He is the author of numerous scientific and popular articles and books, including *Sagebrush Country* (Mountain Press, 1992) and *Northwest Weeds* (Mountain Press, 1990). He is currently writing a book about the wildflowers and ecology of the North American deserts. Taylor lives in Bellingham, Washington.

George W. Douglas has spent nearly 30 years studying plants and plant communities in the mountains of northwestern North America. A Ph.D. botanist, Douglas owns and manages Douglas Ecological Consultants Ltd., an ecological and environmental consulting firm. Since 1991, he has worked as program botanist for the Conservation Data Centre at the British Columbia Ministry of Environment, Lands and Parks. He has written several books, including *Vascular Plants of British Columbia.* Douglas lives in Duncan, British Columbia.